SpringerBriefs in Applied Sciences and Technology

PoliMI SpringerBriefs

More information about this subseries at http://www.springer.com/series/11159
http://www.polimi.it

Alessandro Balducci · Daniele Chiffi ·
Francesco Curci
Editors

Risk and Resilience

Socio-Spatial and Environmental Challenges

Editors
Alessandro Balducci
DAStU
Politecnico di Milano
Milano, Italy

Daniele Chiffi
DAStU
Politecnico di Milano
Milano, Italy

Francesco Curci
DAStU
Politecnico di Milano
Milano, Italy

ISSN 2191-530X   ISSN 2191-5318   (electronic)
SpringerBriefs in Applied Sciences and Technology
ISSN 2282-2577   ISSN 2282-2585   (electronic)
PoliMI SpringerBriefs
ISBN 978-3-030-56066-9   ISBN 978-3-030-56067-6   (eBook)
https://doi.org/10.1007/978-3-030-56067-6

This Springer imprint is published by the registered company Springer Nature Switzerland AG
The registered company address is: Gewerbestrasse 11, 6330 Cham, Switzerland

Introduction

More than 300 natural disaster events were recorded worldwide in 2018, causing about 12,000 deaths and affecting about 68 million people (CRED 2019). Nowadays, 272 million international migrants are causing the depopulation of remote and underdeveloped areas as well as the overpopulation of the most developed cities and regions (IOM 2019). Moreover, income inequalities and the demand for food, energy and water in cities are going to increase dramatically (UN HABITAT 2016).

Starting from these general data and trends, it is possible to put forward the following considerations. On the one hand, climate change is amplifying the number and the magnitude of hazards, and scientific research and specific policies are attempting to prevent, mitigate and respond to environmental risks. On the other hand, the uneven economic development between large urban agglomerations and marginalised rural areas and between central and peripheral urban areas is rapidly increasing social disparities, migration flows and political conflicts.

The concept of resilience is fundamental in facing the challenges posed by both environmental and socio-economic risks. To be effective, resilience requires a highly interdisciplinary attitude and a multi-scalar approach in order to deal with the complexity of socio-spatial systems. Indeed, the interplay between resilience and different forms of risk and uncertainty is crucial in order to handle such complexity and propose innovative policies.

While this book by no means aims to cover all the main topics that have been discussed in the contemporary debate on risk and resilience, we have tried nevertheless to cover some of the most interesting and recently most discussed issues, with a special focus on the Italian context.

The book is intended (i) to discuss the theoretical study of the notions of risk, resilience and allied concepts, (ii) to analyse the current socio-economic trends and phenomena that, at different scales, are making cities and territories vulnerable (Italian ones in particular) and (iii) to offer the main methodological and practical tools operating at the spatial and social levels in order to mitigate risks and increase urban resilience. Indeed, the main objective of this book is to present an accurate and multifaceted approach towards risk and resilience that is sensitive to environmental issues.

In particular, the eight chapters composing the book consider specific approaches aimed at increasing the awareness and skills necessary to face the social, economic and environmental challenges usually encountered in spatial planning. Hence, the book deals with the concepts of risk and resilience from both a theoretical and an operational point of view. All chapters may foster a better understanding of risk and resilience (and other related concepts such as vulnerability, preparedness, fragility, anti-fragility) in urban and landscape studies in order to outline a fine-grained analysis of new planning policies.

Chapter 1, by Daniele Chiffi and Simona Chiodo, provides a critical reflection on the notions of risk and uncertainty and their relevance within an integrated approach to the methodology of planning, encompassing philosophy and urban planning.

Chapter 2, by Alessandro Balducci, points out how planning can contribute to building the resilience of communities and territories in a general context increasingly affected by changes and emergency situations.

Chapter 3, by Scira Menoni, discusses some definitions and interpretations of risk and resilience with the aim of supporting land use and urban planning activities as non-structural mitigation measures.

Chapter 4, by Giovanni Carrosio, deals with the issue of resilience from a socio-territorial point of view and proposes an analytical reading of the socio-ecological crisis, highlighting the interdependencies between the environmental, migration and welfare sustainability issues.

Chapter 5, by Gianfranco Viesti, uses regional and labour systems data to deal with territorial dynamics and disparities in Italy, in particular since the 2008 economic crisis.

Chapter 6, by Gabriele Pasqui, considers the relationship between peripheral conditions and the growth of socio-spatial inequalities in Italian cities and, in general, across Italy. Pasqui proposes a multidimensional approach that links together inequalities, social fragilities and different dimensions of socio-spatial polarisation.

Chapter 7, by Francesco Curci, deals with natural hazard avoidance strategies and describes the most common approaches, which are based on land use regulations and building bans or on managed retreat and relocation programs. Then, the chapter discusses the 'housing decompression' plans and the projects proposed in the last few decades for the hinterland of the Vesuvius volcano.

Chapter 8, by Alessandro Coppola, Silvia Crivello and Wolfgang Haupt, is based on a wide discussion regarding the concept of resilience across global policy mobility networks and focuses on the participation of Milan and Rome in the 100 resilient cities program. The chapter offers a critical analysis of the processes and outcomes of their participation.

The book originates and is promoted within the activities of the research project *Fragilità Territoriali* at the DAStU (Department of Architecture and Urban Studies), Politecnico di Milano (Department of Excellence 2018–2022), in particular, from the PhD course "Approaches to Resilience" from which the project of this book stems.

Alessandro Balducci
Daniele Chiffi
Francesco Curci

References

CRED—Centre for Research on the Epidemiology of Disasters (2019) Natural disasters report 2018. https://reliefweb.int/sites/reliefweb.int/files/resources/CREDNaturalDisaster2018.pdf

IOM—International Organization for Migration (2019) World migration report 2020. IOM, Geneva. https://publications.iom.int/books/world-migration-report-2020

UN HABITAT—United Nations Human Settlements Programme (2016). World cities report 2016: urbanization and development: emerging futures. UN HABITAT, Nairobi. http://wcr.unhabitat.org/wp-content/uploads/2017/02/WCR-2016-Full-Report.pdf

Contents

Chapter 1
Risk and Uncertainty: Foundational Issues

Daniele Chiffi and Simona Chiodo

1.1 Introduction

The aim of this chapter is twofold: first, it is meant to provide its audience (especially non-philosophical readers) with an overview of the philosophical meanings of the notions of risk and uncertainty; and, second, it aims at suggesting which philosophical perspectives are most promising in understanding risks and uncertainties in the complex processes of embrittlement that derive from the relationship between space and society. Moreover, we will explore how our cities and landscapes are exposed to risk and uncertainty because of both natural and human factors. The structure of the present contribution is the following: Sect. 1.1 provides an overview of the philosophical meanings of the notions of risk (Sect. 1.2) and uncertainty (Sect. 1.3); Sect. 1.4 contains a brief suggestion about how philosophers can better collaborate with urban planners and decision-makers working on the possible solutions to the embrittlement of our cities and landscapes; finally, Sect. 1.5 presents our conclusions, showing a unified approach towards a philosophy of urban planning.

1.2 The Philosophical Meanings of the Notion of Risk

Before focusing on the conceptual meaning of the notion of risk, let us start by making reference to its etymological meaning. The etymology of "risk" is quite uncertain: the ancient Italian word *riscare*, which means "to run into danger", may refer to the Latin word *resecare*, i.e. "to cut", or, according to some authors, how rocky outcrops

D. Chiffi (✉) · S. Chiodo
Department of Architecture and Urban Studies, Politecnico di Milano, Milan, Italy
e-mail: daniele.chiffi@polimi.it

S. Chiodo
e-mail: simona.chiodo@polimi.it

cut through ships posing a true navigation hazard. If this makes any sense, then, at least figuratively, we may also think of the process of "cutting" as a kind of division, which means that a given action may have both a positive and a negative outcome since the circumstances make them both possible, and therefore unsure. Indeed, the everyday meaning of the word "risk" seems to refer to circumstances in which it is possible—unlikely as it may be—that a (severely) negative event occurs.

From a historical point of view, the philosophical notion of risk originates from two major cultural revolutions dating back to the seventeenth century: the scientific revolution and the idea according to which one of the most important constitutive qualities in a human being is her/his freedom. This takes us to the necessity of working on the scientific tools we need to manage our freedom, starting from the risks implied by any free action, i.e. by an action whose responsibility is totally ours. Pascal's famous argument on the bet on God's existence, i.e. on the risks implied, may be considered as the first step towards a systematic approach to the notion of risk, and, more specifically, to a contemporary notion of risk assessment. Together with philosophy, other three disciplines have provided seminal tools towards the contemporary definition of the philosophical notion of risk: probability (starting from the seventeenth century), statistics (starting from the nineteenth century) and some of the most recent developments of mathematics (starting from the twentieth century).[1]

Contemporary philosophy may help us understand the complexity of the word "risk", in relation to both its strictly philosophical and its technical/ordinary meanings. The authoritative *Stanford Encyclopaedia of Philosophy* (Hansson 2018) analyses the notion of risk as follows:

(1) risk has five main technical meanings, which deal with both philosophical and non-philosophical application fields;
(2) traditionally, the philosophical application fields of risk have been in six areas, i.e. epistemology, philosophy of science, philosophy of technology, ethics, decision theory and philosophy of economics. As we will see, we believe that some of these disciplines can offer fertile ground to urban studies and be beneficial to urban planners and decision-makers.

1.2.1 Definitions of Risk

In a situation of *certainty,* the possible events are listable, and the consequences of choice are clearly known to the decision-maker; as such, each decision is easily recognised to have a specific outcome. In a situation of *risk* (strictly speaking), the possible events are still listable, and it is possible to specify how likely the event is to occur. Out of the extensive scientific literature, let us consider the classical and technical definition given by the Royal Society (1983), according to which risk is the probability of an event in association with the evaluation of the magnitude

---

[1] A classical text on the history of probability is (Hacking 1975). See also (Morini 2003; 2014).

of consequences in a specific lapse of time.[2] This means that, when uncertainty is expressed in probabilistic terms in accordance with statistical rules and when consequences can be evaluated, we may talk of risk. Still, the notion of risk is quite complex, having both technical and non-technical definitions (Boholm et al. 2016) as well as qualitative (the first and the second below) and quantitative (the third, the fourth and the fifth below) meanings (Hansson 2018):

(1) an unwanted event which may or may not occur;
(2) the cause of an unwanted event which may or may not occur;
(3) the probability of an unwanted event which may or may not occur;
(4) the statistical expectation value of an unwanted event which may or may not occur (where the expectation value is the product of its probability and some measure of its severity);
(5) the fact that a decision is made under conditions of *known probabilities*.

Definition (1) merely stresses the unexpected nature of a risky event, while (2) makes the cause of an event coincide with the risk itself. Of course, it is one thing to talk about risk factors, but quite another to be able to distinguish between risk and causes, which can oftentimes become extremely misleading. Definition (3) highlights the random character of the risky event, regardless of the potential impact of its consequences. Quite the contrary, (4) includes, in the definition of risk, the assessment of possible consequences. As we have seen, definition (4) is used by the Royal Society (1983), where risk is intended in the sense of expected utility. The last definition (5) stresses how decisions taken under risk conditions fall within the scope of *known unknowns*, that is, of those events that may or may not occur and of whose potential occurrence we have at least a probabilistic assessment. In the field of disaster risk assessment, in particular, the following are identified as risk components: the *potential danger (hazard), the exposed value* (or *exposure*), and the *vulnerability*, which can be defined as the susceptibility of the exposed elements (people, manufactured products, economic activities, etc.) to suffer damage caused by a specific potentially harmful event (UNISDR 2015).

The distinction between natural and human (esp. technological) risks, which has been commonly pointed out in literature, turned out to be useful in many contexts but does not seem to be exhaustive. In fact, some risks can be seen as both natural and related to human actions. For this reason, it is more convenient to try to establish the proportion of a natural risk and that of a human-induced one, so as to be able to understand how these two types of risk interact with each other. This means that in risk analysis human and non-human factors are deeply interrelated.

In some scientific fields, including planning and economics, experts, rather than referring to risk, may refer to the narrower notion of *parametric uncertainty*, meaning that the possible events are still in some way listable, but the decision-maker lacks to know *ex ante* the values of some parameters of a process that may come to be known only *ex post*. In short, it is a situation of epistemic uncertainty in which there is "a

---

[2]For an analysis of the fundamental elements of risk theory, see Roeser et al. (2012).

lack of complete knowledge *ex ante* about the values that specific variables within a given problem structure will take *ex post*" (Langlois 1994, 118).

1.2.2 The Philosophical Applications of the Notion of Risk

The contemporary philosophical debate on the notion of risk and its application fields is rich, and this includes the criticism against its above-mentioned technical meanings. The six philosophical areas to which it has been (and is still being) applied are the following.

(1) *Epistemology*: according to this knowledge-focused philosophical discipline, risk occurs when there is lack of knowledge, making an event uncertain. Yet, risk and uncertainty are different notions (as we shall see better in Sect. 2) even if very often they are used as synonyms, the former being more objective (and quantitatively identifiable, at least in most cases) than the latter. This is due to the fact that the notion of risk makes reference to an event whose relevant probabilities are available (or at least idealizable by a model, and therefore somehow predictable), whereas the notion of uncertainty makes reference to an event whose relevant probabilities are partially or completely unavailable. Of course, in real life the first case is less frequent than the second, because of the extreme complexity of many of the fields in which we try to make predictions, from natural to social events (Shrader-Frechette 1997). Moreover, epistemology makes another important distinction by dividing objective risks from subjective risks, where the latter imply a subjective estimate of the former, based on a variety of subjective factors, from perceptions to expectations (Sjöberg 2000).

(2) *Philosophy of science*: the main issues are, on the one hand, the role of values in scientific fields dealing with risks as well as in risk assessment, and, on the other hand, the relationship between scientific evidence, epistemic risks of error and policy decision.[3] As regarding the first case, in risk assessment the tools of philosophy of science can help us identify two kinds of errors we may make: false positives, whenever we think there is a risk that does not actually occur, and false negatives, whenever we think there is no risk but it actually occurs. As for the second case, the issue at stake is how the risk of error in science, especially in falsely accepting or rejecting hypotheses, may influence practical decisions in general and policy decisions in particular (Hempel 1965; Cranor 1990; Douglas 2000; Chiffi and Giaretta 2014). In fact, the increasing amount of disinformation fomented by the fake news phenomenon is having a real public backlash on policy-making (Cranor 2016).

(3) *Philosophy of technology*: one of the core topics is safety engineering (Möller and Hansson 2008, Doorn and Hansson 2015). In particular, three issues are analysed: *inherent safety* (i.e. elimination of hazards), *safety factors* (i.e. the

---

[3]On the role of public values and the need for interdisciplinary research in science and policy-making, see (Taebi et al. 2014).

design of technologies that are stronger than the minimum required) and *multiple barriers* (i.e. strictly independent security systems designed to come into play if the predecessor fails).

(4) *Ethics*: the focus is here on the moral aspects of risk-taking, in particular on the role played by intentionality and will of both subjects of decisions (risk-taking) and objects of decisions (risk-imposing). These matters can be analysed through the standard moral theories: *utilitarianism* (a form of consequentialism in which actions are judged by the amount of pleasure and pain they cause: the action that brings the greatest happiness for the greatest number should be chosen; see Scanlon 1982), *rights-based moral theories* (see, in particular, Nozick 1974), *deontological moral theories* (here, analogously to the previous case, the challenge is where to set the threshold between reasonable and unreasonable impositions of risks based on moral rules and principles) and finally *contract theories* (grounded on the criterion of consent among all the individuals involved as well as on strategies of thought experiment and idealisation; see Rawls 1971).

(5) *Decision theory*: its aim is the identification of how to reach the best possible outcome, given a certain set of values, whenever we act under risk. In the next sections of our contribution, we will more specifically focus on planning decisions under conditions of risk and of uncertainty.

(6) *Philosophy of economics*: risk plays a major role in economics, and having a deeper understanding of the philosophical notion of risk can help dealing with risk management pitfalls. In particular, the issues at stake are risk attitude measures (often subjective and non-rational) and the use of risk-benefit (or cost-benefit) analysis in order to weigh out the advantages and disadvantages of a decision in numerical, i.e. monetary terms (Sunstein 2005). In this particular case, the philosophical challenge has generally to do with the very possibility of assigning a monetary value to a human life, leading eventually to its loss. The extreme complexity of some scenarios, e.g. risks of social injustice, can make its application quite difficult.

Another relevant philosophical application field of the notion of risk may be aesthetics. Although this new entry may come off as rather unconventional, it is bound to play a promising role. On the one hand, aesthetics may offer fundamental insights to some of the above-mentioned issues, such as the distinction between objective and subjective risks, whereby the latter involves perceptions (a traditional aesthetics' topic) and expectations (often driven by emotions, which are another traditional aesthetics' topic). On the other hand, aesthetics may open up other philosophical application fields by analysing, for instance, the role played by artefacts and works of art on subjective risks. Consider, for instance, a pack of cigarettes: the obscene images of human bodies suffering from severe diseases show just how powerful aesthetic tools can be when compared to scientific tools in making people understand, and possibly prevent, risks. Thus, they can assume a pivotal role in policy decision.

1.3   The Philosophical Meanings of the Notion of Uncertainty

Let us start, once again, by considering the etymology of the word "uncertainty". The term is based on the negation of the adjective "certain" deriving from the Latin adjective *certus*, and therefore from the Latin verb *cernere*, which literally means "to sift" (in particular, by separating the flour from the bran) and, figuratively, "to distinguish"; more precisely, "to distinguish" the true from the false. Thus, the word "uncertain" means that there is something, be it a judgment or an event, in which we cannot "distinguish" what is true from what is false, or also what will (truly) occur from what will not (truly) occur. Uncertainty, in this sense, refers to a lack of knowledge and, more specifically, to situations in which we know the type of consequences but cannot meaningfully attribute probabilities to the occurrence of the event entailing such consequences. When even the events are unknown, it is standard to talk of ignorance (Carrara et al. 2019).[4] Quite interestingly, many realistic decision-making processes may face extreme forms of uncertainty. Fundamental (severe, genuine, great, extreme) uncertainty exemplifies common forms of uncertainty for which it is extremely difficult to formulate meaningful probabilistic evaluations. As observed by Keynes:

> By "uncertain" knowledge, let me explain, I do not mean merely to distinguish what is known for certain from what is only probable. The game of roulette is not subject, in this sense, to uncertainty [...]. The sense in which I am using the term is that in which the prospect of a European war is uncertain, or the price of copper and the rate of interest twenty years hence [...]. About these matters there is no scientific basis on which to form any calculable probability whatever. We simply do not know. (Keynes 1979: 113–114)[5]

Even in presence of fundamental uncertainty, some argumentative and methodological tools have been proposed in order to mitigate the impact of extremely uncertain events (that may convey great potential losses; Chiffi and Pietarinen 2017). A common way to approach this type of questions consists in an appeal to the precautionary principle, which prescribes how to deal with threats that are uncertain and/or cannot be scientifically established. A general version of the precautionary principle has the following form: "If there is (i) a threat, which is (ii) uncertain, then (iii) some kind of action (iv) is mandatory" (Van de Poel and Royakkers 2011). Without entering the vast literature on this principle, it is undeniable that in some circumstances the adoption of the precautionary principle in its general form may result in stopping innovation. For this reason, other approaches towards decision and argumentation under fundamental uncertainty have been developed. Some indications taken by Hansson (1996, 2011) may help us justify our decisions under fundamental uncertainty: (1) unlike standard scenario planning, a search for mere *possibility arguments* pointing in different directions should be adopted. The idea here is to extend

---

[4] A semantic notion related to uncertainty is ambiguity, i.e. the property that different interpretations or meanings can be given to a same? term (Dequech 2000). On this point, philosophers may contribute to the conceptual clarifications of terms and notions.

[5] See also (Knight 1965).

the range of scenarios to be considered in decision-making by taking into account the ones that are possible even if not fully plausible. This is particularly true when possible scenarios are equipped with potentially severe consequences. (2) A *scientific evaluation* that may lead to the specification or refutation of some of these arguments. This means that scenarios, when possible, should be evaluated trying to apply methods, models and tools of scientific reasoning. In some cases, in fact, uncertainty may be due to scientific methodology. (3) We may imagine *a symmetry test for causes and effects*, i.e. we can maintain the cause(s) of an event fixed and then try to conceive potential consequences in different scenarios. Conversely, we may keep consequence(s) fixed and try to retrospectively isolate some potential causes. (4) We need to make an evaluation of the *seriousness of the considered scenario* in terms of the *novelty, spatio-temporal unlimitedness* and *interference with complex systems* of the threats involved. When we consider a hazard that is new in a given scenario and may spread without limitations in space and time thus possibly interfering with other complex systems, we can infer that such scenario may entail severe potential consequences. (5) *Hypothetical retrospection* is an argumentative methodology that aims at finding a course of action that will be defensible in retrospect. According to hypothetical retrospection, decisions are evaluated (in particular, from a moral perspective) assuming that a possible branch of future developments has materialized. Such evaluation is based on the values and information available when the original action took place, from the point of view of the imagined future point of retrospection. The decision rule for retrospective judgement requires choosing an alternative that emerges as (morally) acceptable from all hypothetical retrospections. This methodology assumes that time has a branching structure and relies on the role of values that existed when the decision took place. This means, though, that hypothetical retrospection may not properly be applied in uncertain scenarios in which there is a rapid change of values.

1.4 From Philosophy to the Emergencies of the Space We Live In

Our overview of the philosophical meanings of the notions of risk and uncertainty allows us to show some of their potential uses when it comes to facing risks and uncertainties related to the space we live in. Philosophy, in fact, can help us find new ways of facing the complex processes of embrittlement deriving from the relationship between space and society, also because it contributes to identifying the different forms of risk and uncertainty our cities and landscapes are exposed to.

1.4.1 Risks of the Space We Live In

Imagine a city where there is a big factory that employs many of the citizens and considerably contributes to its richness. At a certain point, several of the employees of the factory start getting ill, and some of them even die. The dramatic circumstance is carefully studied by some scientists, who state that, even if there are some clues pointing to a possible correlation between the factory and the casualties, such evidence cannot be taken as scientific. Thus, on one side, we have some clues of a deadly risk and, on the other, we have no actual scientific evidence for it. What would you do if you were the mayor of the city? In this case, the epistemological distinction between objective and subjective risks can help you identify and manage two kinds of dimensions which should be distinguished. The first dimension has to do with the objective risk of getting ill and even die: in order to identify it as precisely as possible, you will need to both refer to more than one scientific study (since the comparison of these results is always methodologically helpful) and to exclude scientists whose judgments may be influenced by subjective evaluations of risks (since they may be, for instance, the relatives of one of the deceased, or of one of the owners of the factory). The second dimension, instead, has to do with the subjective risk of getting ill and even die: while you are identifying the objective risk as precisely as possible, you should adopt a behaviour that helps avoid the two possible extremes caused by your citizens' subjective risk, i.e. either totally ignoring it (and acting accordingly) or totally dramatizing it (and acting accordingly). Possessing a philosophical awareness of these kinds of distinctions as well as of their causes and effects can help you act in a rational way, by finding a balance between underestimation and overestimation of both objective and subjective risks.

Having a philosophical awareness can also help you in the case of risk assessment, even in its most complex occurrences. Imagine, for instance, that you are the mayor of a multi-ethnic city, where a representative of the Islamic community asks you to build a mosque, whereas a representative of the autochthon community is against such idea. The former's argument is the risk of terrorism if the Islamic community is not treated in an inclusive way, whereas the argument of the latter is the risk of losing the autochthon religious identity if the mosque were to be built. Again, what would you do if you were the mayor of the city? In similar situations, philosophy of science may be helpful in assessing the risk by making explicit the values that underlie the opposite perceptions. In the first case, you may identify and make explicit a value which seems to be helpful when it comes to managing a multi-ethnic city: the value of peaceful cohabitation (indeed, facts seem to support the correlation between social ghettoization and risk of terrorism). In the second case, you may identify and make explicit a value which seems, instead, to be unhelpful in managing a multi-ethnic city: the value of nationalism (indeed, facts do not seem to support the correlation between social inclusion and risk of losing the autochthon religious identity: building a mosque does not mean depriving Christians from praying in churches). Thus, making values explicit can even help us distinguish between "real" and "unreal" risks.

The above-mentioned examples imply another philosophical field of application that turns out to be crucial for us to face risks: ethics, which analyses our ways of acting, starting from possible reasons and outcomes. The examples we used, moreover, shed light on one of the main issues ethics focuses on: our actions, which follow our decisions, are likely going to influence others. Thus, our responsibility increases, and managing it becomes one of the most complex matters of our lives. Ethics can help us identify our intentionality. In the example of the mayor, we should be able to identify, and remove, any possible interference between what is advantageous for ourselves and what is advantageous for our citizens. But ethics can also help us distinguish and evaluate the consequences of risk-imposing and risk-taking decisions. Surely, the former situation stands out as more challenging than the latter because in that case we are not only deciding under risk, but also under risk for others' lives. Such distinction should be clear if we were the city mayor. Yet, we should be able to recognize this "threat" even in more ambiguous circumstances, such as, for example, those in which architects happen to choose a certain eccentric formal solution for a construction instead of carefully considering its impact on others' lives.

Moreover, we should not forget that the decision to build new infrastructure and artefacts is also justified by economic reasons. Cost–benefit analysis (CBA), as we have seen, is an economic and consequentialist method of evaluation in which uncertainty is assumed as (probabilistic) risk. CBA is a monetary valuation of different effects of interventions that is undertaken by using prices revealed by the market or by agents' "willingness to pay for (or accept compensation to avoid) different outcomes" (Drummond et al. 2005). Some philosophical and theoretical limitations of CBA have been identified in the literature (Hansson 2007; Sunstein 2005), namely: (1) values may be implemented in CBAs to mitigate some anomalies; (2) the problem of incommensurability of values and monetization of the outcome of different decisions; (3) the possible disagreement on the issue of the common (monetary) scale for values; (4) the stakeholders' values and different perspectives may modify the results of a CBA; (5) CBA does not always guarantee a just distribution of costs and benefits; and finally (6) some forms of CBA should require informed consent.[6]

The methodology of CBA is strictly connected with risk analysis. The idea is that CBA can have something to say when it is directed towards the construction of a standard infrastructure like, for instance, a new bridge. In this case, in fact, the scenarios associated to the decision are easily listable; and although some predictions are required (such as, for instance, the expected number of people using the new

---

[6]Cost-benefit analysis also finds a nice application in what is called "Economy of Research" by C. S. Peirce, i.e. a methodology to evaluate the plausibility and feasibility of scientific hypotheses. Peirce (1931–58) pointed out that "now economy, in general, depends upon three kinds of factors: cost; the value of the thing proposed, in itself; and its effect upon other projects. Under the head of cost, if a hypothesis can be put to the test of experiment with very little expense of any kind, that should be regarded as a recommendation for giving it precedence in the inductive procedure" (CP 7.220). On the Economy of Research, see: (Rescher 1976; Haack 2018; Tuzet 2018; Chiffi et al. 2020). On the role of values and economic considerations in shaping science, see (Chiffi and Pietarinen 2019).

infrastructure), the impact on space and time of the new infrastructure is usually limited and well-defined (Moroni and Chiffi 2021). Thus, in these kinds of situations, scenarios and consequences may be easily conceived and evaluated.

1.4.2 Uncertainties of the Space We Live In

We hold that cities are the canonical example of uncertainties permeating the space we live in. Even if some common trends have been acknowledged in the study of cities (Bettencourt and West 2010), the individual behaviour of people living in urban spaces may be unpredictable and deeply uncertain. Although urban uncertainty is becoming the focus of an increasing number of scientific papers in the literature on urban planning, this issue would certainly require further (interdisciplinary) analysis.[7] Planning under uncertainty may be considered *prima facie* as contradictory. However, uncertainty permeates all concrete cases of (urban) planning.

Cities are fundamentally uncertain[8] because (1) the potential scenarios associated to their development are hardly listable, (2) the consequences of urban planning decisions may result in completely unpredictable alternatives since reliable estimates are rarely possible, (3) cities are complex systems and highly interconnected so that they may have a great and unlimited interference in space and time as well as with other complex economic and social systems. Usually, a simple appeal to pragmatic and adaptive strategies in urban planning has been advocated in the literature in order to deal with uncertainty (Kato and Ahern 2008; Skrimizea et al. 2019). We think, nevertheless, that a theoretical understanding of uncertainty in urban systems may contribute to highlighting the multifaceted complexity of current cities. Many urban uncertainties, in our age of big data, are treated as urban risks exemplified by a number of statistics, frequencies and probabilistic assessments. Still, this attitude seems to conflate conditions of uncertainty with conditions of risk in the urban domain. This does not mean that a quantified approach towards relevant urban factors is meaningless; but embracing the complexity of our cities needs us to open up the range of futures in order to deal not just with plausible scenarios but also with possible and unexpected ones. Knowledge of the past and knowledge of trends do not always guarantee the possibility of accurate predictions, which represents, instead, a key factor in urban planning. Moreover, fundamental uncertainty in planning is classically connected with the so-called "wicked" problems (Rittel and Webber 1973; Hajer, Hoppe and Jennings 1993; Balducci et al. 2011)[9]; that is, those problems that are difficult to be consistently and sharply formulated since their understanding and their

---

[7]See e.g. Abbott (2005, 2009), Beauregard (2018), Kato and Ahern (2008), Nyseth (2012), Savini (2017), Skrimizea et al. (2019), Spirandelli et al. (2016), Stults and Larsen (2018), Zapata and Kaza 2015, Zandvoort et al. (2018).

[8]Urban complexity is another key feature needed to investigate contemporary cities (Moroni and Cozzolino 2019).

[9]See also Balducci's chapter in this book.

resolution are concomitant to one another. This happens because, in order to anticipate the questions related to these kinds of problems, knowledge of all potential solutions is required. Given that "wicked" problems come in a complex form and are often ill-defined, all planning problems seem to be basically wicked. The methodologies involved in the mitigation of fundamental uncertainty may provide us with the proper tools to deal with wicked problems within this context.

1.5   Conclusion

The present contribution has served to outline the distinctive elements of the notions of risk and uncertainty. We have thoroughly discussed a wide range of technical and non-technical definitions of risk from a philosophical perspective, and we have briefly touched upon some of their standard applications in risk theory. Then, we have introduced the notion of uncertainty and provided some methodological suggestions on how to deal with it. On a last note, we have made a clear differentiation between planning scenarios involving risk and those involving uncertainty. Since cities are fundamentally uncertain, breaking down these notions may be problematic for urban planning. This is why the philosophy of urban planning may become a promising interdisciplinary field at the crossroads between disciplines with many potential applications.[10]

Acknowledgements This research is supported by the Excellence Project *Fragilità Territoriali* (2018–2022; L. 232/2016) of the Department of Architecture and Urban Studies (DAStU; Politecnico di Milano). We thank all the colleagues at DAStU for their suggestions and observations.

References

Abbott J (2005) Understanding and managing the unknown: the nature of uncertainty in planning. J Plann Educ Res 24(3):237–251. https://doi.org/10.1177/0739456X04267710
Abbott J (2009) Planning for complex metropolitan regions: a better future or a more certain one? J Plann Educ Res 28(4):503–517. https://doi.org/10.1177/0739456X08330976
Albrechts L, Balducci A, Hillier J (eds) (2016) Situated practices of strategic planning: an international perspective. Routledge, London
Balducci A, Boelens L, Hillier J, Nyseth T, Wilkinson C (2011) Introduction: strategic spatial planning in uncertainty: theory and exploratory practice. Town Plann Rev 82(5):481–501. https://doi.org/10.3828/tpr.2011.29
Beauregard R (2018) The entanglements of uncertainty. J Plann Educ Res. https://doi.org/10.1177/0739456X18783038
Bettencourt L, West G (2010) A unified theory of urban living. Nature 467(7318):912–913. https://doi.org/10.1038/467912a

---

[10] A strategic planning approach as a collective effort to reimagine cities, urban regions and lands (Albrechts et al. 2016) may be the proper ground for this interdisciplinary challenge.

Boholm M, Möller N, Hansson SO (2016) The concepts of risk, safety, and security application in everyday language. Risk Anal 36(2):320–338. https://doi.org/10.1111/risa.12464

Carrara M, Chiffi D, Florio De C, Pietarinen A-V (2019) We don't know we don't know: asserting ignorance. Synthese 1–16.https://doi.org/10.1007/s11229-019-02300-y

Chiffi D, Giaretta P (2014) Normative facets of risk. Epistemologia 37:22–38. https://doi.org/10.3280/EPIS2014-002003

Chiffi D, Pietarinen A-V (2017) Fundamental uncertainty and values. Philosophia 45(3):1027–1037. https://doi.org/10.1007/s11406-017-9865-5

Chiffi D, Pietarinen A-V (2019) Risk and values in science: a Peircean view. Axiomathes 29(4):329–346. https://doi.org/10.1007/s10516-019-09419-0

Chiffi D, Pietarinen A-V, Proover M (2020) Anticipation, abduction and the economy of research: the normative stance. Futures 115:102471. https://doi.org/10.1016/j.futures.2019.102471

Cranor CF (1990) Some moral issues in risk assessment. Ethics 101(1):223–243. https://doi.org/10.1086/293263

Doorn N, Hansson SO (2015) Design for the value of safety. In: van den Hoven J, van de Poel I, Vermaas P (eds) Handbook of ethics, values and technological design. Springer, Dordrecht, pp 491–511

Dequech D (2000) Fundamental uncertainty and ambiguity. East Econ J 26(1):41–60. https://www.jstor.org/stable/40325967

Douglas H (2000) Inductive risk and values in science. Philos Sci 67(4):559–579. https://doi.org/10.1086/392855

Drummond MF, Sculpher MJ, Torrance GW, O'Brien BJ, Stoddart GL (2005) Economic evaluation in health care: merging theory with practice. Oxford University Press, Oxford

Haack S (2018) Expediting inquiry: Peirce's social economy of research. Transactions of the Charles S. Peirce Society 54(2):208–230. https://www.jstor.org/stable/10.2979/trancharpeirsoc.54.2.05

Hacking J (1975) The emergence of probability. Cambridge University Press, Cambridge

Hajer MA, Hoppe R, Jennings B (1993) The argumentative turn in policy analysis and planning. Duke University Press, Durham, NC

Hansson SO (1996) Decision making under great uncertainty. Philos Soc Sci 26(3):369–386. https://doi.org/10.1177/004839319602600304

Hansson SO (2007) Philosophical problems in cost-benefit analysis. Econ Philos 23(2):163–183. https://doi.org/10.1017/S0266267107001356

Hansson SO (2011) Coping with the unpredictable effects of future technologies. Philos Technol 24(2):137–149. https://doi.org/10.1007/s13347-011-0014-y

Hansson SO (2018) Risk, The Stanford Encyclopedia of Philosophy. In: Zalta E (ed). https://plato.stanford.edu/archives/fall2018/entries/risk/

Hempel CG (1965) Aspects of scientific explanation and other essays in the philosophy of science. The Free Press, New York

Kato S, Ahern J (2008) 'Learning by doing': adaptive planning as a strategy to address uncertainty in planning. J Environ Planning Manage 51(4):543–559. https://doi.org/10.1080/09640560802117028

Keynes JM (1979) The general theory and after: defence and development. In: The Collected Writings of John Maynard Keynes, vol XIV. Macmillan, London

Knight FH (1965) Risk uncertainty and profit. Harper & Row, New York

Langlois RN (1994) Risk and uncertainty. In: Boettke PJ (ed) The Elgar companion to Austrian economics. Edward Elgar, Cheltenham, pp 118–122

Morini S (2003) Probabilismo. Storia e teoria. Bruno Mondadori, Milan

Morini S (2014) Il rischio. Da Pascal a Fukushima. Bollati Boringhieri, Turin

Moroni S, Cozzolino S (2019) Action and the city: emergence, complexity, planning. Cities 90:42–51. https://doi.org/10.1016/j.cities.2019.01.039

Moroni S, Chiffi D (2021) Complexity and uncertainty: implications for urban planning. In: Portugali J (ed) Handbook on cities and complexity. Edward Elgar Publishing, Cheltenham, UK

Nyseth T (2012) Fluid planning: a meaningless concept or a rational response to uncertainty in urban planning? In: Burian J (ed) Advances in spatial planning. InTech, Rijeka, pp 27–46

Nozick R (1974) Anarchy, state, and utopia. Basic Books, New York

Peirce CS (1931–58) Collected papers. In: Peirce CS, Weiss P, Burks A (eds) Harvard University Press. Cambridge, MA (References are to CP by volume and paragraph number)

Rawls J (1971) A theory of justice. Harvard University Press, Cambridge

Rescher N (1976) Peirce and the economy of research. Philos Sci 43(1):71–98. https://doi.org/10.1086/288670

Rittel HW, Webber MM (1973) Dilemmas in a general theory of planning. Policy Sci 4(2):155–169. https://doi.org/10.1007/BF01405730

Roeser S, Hillerbrand R, Sandin P, Peterson M (eds) (2012) Essentials of risk theory. Springer Science & Business Media, Cham

Royal Society (1983) Risk assessment: report of a royal society study group. Royal Society, London

Savini F (2017) Planning, uncertainty and risk: the neoliberal logics of Amsterdam urbanism. Environ Plann A 49(4):857–875. https://doi.org/10.1177/0308518X16684520

Scanlon TM (1982) Contractualism and utilitarianism. In: Sen A, Williams B (eds) Utilitarianism and beyond. Cambridge University Press, Cambridge, pp 103–128

Shrader-Frechette K (1997) Hydrogeology and framing questions having policy consequences. Philos Sci 64:S149–S160. https://doi.org/10.1086/392595

Sjöberg L (2000) Factors in risk perception. Risk Anal 20(1):1–11. https://doi.org/10.1111/0272-4332.00001

Skrimizea E, Haniotou H, Parra C (2019) On the "complexity turn" in planning: an adaptive rationale to navigate spaces and times of uncertainty. Plann Theor 18(1):122–142. https://doi.org/10.1177/1473095218780515

Spirandelli DJ, Anderson TR, Porro R, Fletcher CH (2016) Improving adaptation planning for future sea-level rise: understanding uncertainty and risks using a probability-based shoreline model. J Plann Educ Res 36(3):290–303. https://doi.org/10.1177/0739456X16657160

Stults M, Larsen L (2018) Tackling uncertainty in US local climate adaptation planning. J Plann Educ Res.https://doi.org/10.1177/0739456X18769134

Sunstein CR (2005) Cost-benefit analysis and the environment. Ethics 115:351–385. https://doi.org/10.1086/426308

Taebi B, Correlje A, Cuppen E, Dignum M, Pesch U (2014) Responsible innovation as an endorsement of public values: the need for interdisciplinary research. J Respons Innov 1(1):118–124. https://doi.org/10.1080/23299460.2014.882072

Tuzet G (2018) On "The Economy of Research". Trans Charles S. Peirce Society 54(2):129–133. https://www.jstor.org/stable/10.2979/trancharpeirsoc.54.2.01

United Nations Office for Disaster Risk Reduction (UNISDR) (2015) UNISDR Annual Report 2015: 2014–15 Biennium Work Programme Final Report, Geneva. https://www.unisdr.org/files/48588_unisdrannualreport2015evs.pdf

Van de Poel I, Royakkers L (2011) Ethics, technology, and engineering: an introduction. Wiley, Malden, MA

Zapata MA, Kaza N (2015) Radical uncertainty: scenario planning for futures. Environ Plann B Urban Anal City Sci 42(4):754–770. https://doi.org/10.1068/b39059

Chapter 2
Planning for Resilience

Alessandro Balducci

2.1 The Loss of the State of Stability

In his 1971 book, Donald Schön addresses the growing turbulence and instability of the economic and social framework. In "Beyond the Stable State", he asserts that the entire history of humanity can be interpreted as the alternation of long periods of stability with short periods of change, crisis or revolution. The public policy framework, he says, has always been linked to this condition, and in particular to long periods of stability with the occasional need for adaptation linked to rare occurrences of turbulence. The paradigm of Technical Rationality is based on this concept: if problems are stable and recurrent, they can be analysed and catalogued, just as it is possible to develop a repertoire of solutions applicable to that catalogue. But this condition no longer exists, says Schön; the state of stability has been irreparably lost.

This text was written in the early 70s, when many researchers began to indicate that a turning point in public policy had been reached. The 1970s saw the first vigorous expansion of environmental movements, some of the main achievements of the welfare state, the oil crisis of 1972–73, and the economic consequences thereof that continued throughout the decade, precipitating the great neo-liberal turning point inaugurated by the elections of Margaret Thatcher in Great Britain (1979) and Ronald Reagan in the United States (1981).

Indeed, the 70s are considered a crucial period in much of the literature, including by researchers from the Collective for Fundamental Economy (2018). They distinguish between services of the fundamental "material" economy, consisting of basic infrastructure (electrification, aqueducts, sewers, telephone networks, food distribution networks), and the "providential" fundamental economy, consisting of welfare services (education, health, and social safety nets). The main impetus for the former

A. Balducci (✉)
Department of Architecture and Urban Studies, Politecnico di Milano, Milan, Italy
e-mail: sandro.balducci@polimi.it

came in the second half of the nineteenth century, while the latter saw extraordinary development in the second post-war period. The authors' point is that both these essential services sectors, which have been fundamental since the 1970s, have undergone privatization and contraction processes that have penalized the weaker sections of the population, whose well-being is strongly linked to the affordability and efficiency of such services.

The 70s also saw the end of the "glorious 30", the period beginning in the immediate post-war period and ending with the first oil crisis, which forced the world to grapple for the first time with the limits of development, international competition, and globalization.

From then on, a series of phenomena unfolded that I can mention only briefly:

- After the great urbanization of the 1960s, many cities in the West began to lose population to neighbouring territories while cities expanded into surrounding territories; there was talk of an urban crisis (Ceccarelli 1978) with parallel growth of urban sprawl (Indovina 1990) and secondary poles.
- Urbanization processes accelerated significantly all over the world, giving rise to a variety of urban forms, from the megacities of Africa, Asia, and Latin America (Burdett and Sudjic 2007) to the megacity regions of the West (Hall and Pain 2006).
- Consequently, greenhouse gas emissions increased in an uncontrolled way, as indicated by the reports of the Intergovernmental Panel on Climate Change that have been regularly published since 1990 (IPCC 1990).
- With the rise of the Internet, the last decades of the 1900s saw an acceleration of global interconnection and competition between cities, with a return to the concentration of investments in large cities after the urban crises of the 70s and 80s (Florida 2008).
- Internal areas were gradually abandoned; these are territories that were marginalized in the logic of global competition; abandonment involves lack of oversight, disinvestment in the territory, and a decrease in the supply of services (De Rossi 2018).
- Neo-liberal choices in many countries of the more developed world redirected public spending according to "rationalization" criteria that increasingly removed basic equipment from weak areas: hospitals, courts, high schools, and branches of low-traffic railways were closed, and public housing was almost completely blocked.
- With the new century, the effects of progressive global warming become more marked: drought, hurricanes, floods, landslides, and devastating fires, accompanied in some countries by destructive earthquakes, including in Italy.
- After 11 September 2001, with the attack on the World Trade Centre, terrorist attacks became the constant of a previously unknown destructive phenomenology with victims all over the world in the 2000s, from the Near East to the heart of western countries.
- The progress of globalization, while making it possible to relieve significant portions of the world population of hunger, produced a marked increase in

inequalities in the most advanced countries; this phenomenon also assumes strong characteristics of territorial inequality.

- Social and territorial polarization has produced manifest and latent tensions, which have found their way into a series of movements and populist parties that have imposed themselves in national elections, as in Italy, Brexit, and Trump's election, resulting in protectionist policies that have undermined the strengths of those countries; Rodríguez-Pose (2018) describes this phenomenon as "the revenge of the territories that don't matter".
- Latour (2018) also interprets increasingly frequent environmental disasters, from hurricanes to the melting of glaciers and the consequent rise of the oceans, as nature's revenge against abuse. With Donald Trump's withdrawal from the COP 21 Paris Agreement amidst the looming environmental crisis, Latour finds that the richest part of the West has cynically ascertained that we cannot save everyone, and thus with the motto "America First" the United States has decided to ignore common environmental efforts by unloading the effects of its unsustainable choices on the rest of the world.
- In the 2000s, many countries in the North of the world received vast migrations from the poorest countries afflicted by war, hunger and famine; viewed by right-wing parties as an invasion, this crisis has seen a trampling of the basic principles of solidarity and human rights, with deaths in the Mediterranean and on the border between the United States and Mexico, as huge numbers of people flee to refugee camps at the borders of the richest areas.
- In recent months, the revolt of the Yellow Vests has turned France upside down, triggered by a protest against increased taxation on fuel, which had the purpose of protecting the environment.
- Finally, the most serious crisis that the world has been experiencing for some months: the COVID-19 pandemic, has spread rapidly throughout the planet, fuelled by the extension of urbanization processes, unfavourable environmental conditions, and by the hyper-mobility of a population that now lives globalization.

All this produces great instability, with effects on society, economy, and space.

Donald Schön's prediction of the loss of the state of stability has finally come true, and the forms of planning are progressively driven to question their own rules and the role they can play in an increasingly unstable society threatened by social and environmental problems.

2.2 The Contribution of Theoretical Reflection on Planning

It must be said that, for a long time, the shrewdest reflections on the canon and forms of planning has severely and rightly criticized the inability of the dominant planning paradigm to deal with increasing complexity.

Rittel and Webber (1973) article, perhaps one of the most cited in the planning literature, affirms that the planning problems with which contemporary society is confronted are "wicked" problems, and very different from the "benign" problems of science, math or engineering, which they consider "tame". The dominant planning paradigm is built on the criteria of technical rationality, and is effective only for stable situations, for benign problems.

Planning problems are wicked because they do not have a definitive formulation; their potential for treatment depends on how they are formulated. They are substantially unique; it is not possible to catalogue them as with benign problems. Their solution is not true or false, but good or bad depending on the point of view of the judge. Wicked problems are never completely solved, only variously attacked. Unlike the problems of the natural sciences, there is no rule to determine when to stop efforts, nor do we have scientific criteria that allow us to tell when a wicked problem has been solved.

Much of the difficulty therefore arises from the different dimensions of uncertainty that characterize planning problems, both with reference to objectives and intervention technologies.

Schön himself, in another important text entitled "The Reflective Practitioner" (Schön 1983), strongly affirms that the uniqueness of the problems in the turbulent situations in which planners operate renders obsolete the Technical Rationality by which planning had long been inspired. If problems are substantially unique, a new epistemology of professional practice is needed, based on the ability to construct an understanding of the specific characteristics of a given problem, through dialogue among different sources of expertise and mobilised actors.

Planning that has its roots in technical rationality is static and unable to cope with change. To deploy its capabilities, it needs a state of stability that either no longer exists or has been greatly reduced. The choice is between rigor and relevance: one can stay in the high lands where standardized methodologies can be rigorously applied but there are few problems, or in the swamp, where most of the problems are, and where a complex attempt to deal with them can be really relevant.

Karen Christensen, a pupil of Melvin Webber, sought in a well-known essay to articulate the reasoning around the uncertainty and uniqueness of problems (Christensen 1985) and advocated freedom from the constraints imposed by the rational paradigm in order to open planning up to more adaptive forms, whose character must necessarily depend on the problematic situation faced: it is important to get away from a logic in which consensus on objectives and knowledge of intervention technologies are presumed.

Combining these two factors in a well-known chart (Fig. 2.1), she affirms that only a few urban and social development problems are benign and lack uncertainty about the means and ends of the action. Only in those cases are the methods and forms of traditional planning applicable. When there is more uncertainty regarding intervention technologies than consensus on objectives, experimentation is necessary, and vice versa when uncertainty concerns above all objectives rather than intervention technologies, one must embrace bargaining; when uncertainty concerns consensus on objectives, solutions, and intervention technologies, one must work on the complex process of re-framing and redefining problems.

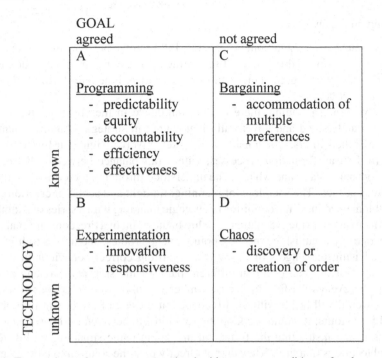

Fig. 2.1 Expectations of government associated with prototype conditions of and responses to planning problems (From Christensen 1985)

The key, according to Christensen, is therefore to abandon rigid and only seemingly reassuring forms of planning, and to assume an adaptive, procedural and, as many authors begin to define it, strategic concept.

Strategic planning also began to be adopted following failures of traditional forms of planning in the second half of the 80s. Planners in the United States and Europe were influenced by the success of strategic planning in private enterprise as well as the growing difficulties in urban planning due to attacks by neoliberal governments (Bryson and Roering 1987; Albrechts et al. 2003). Strategic planning breaks with the static nature and "separateness" of urban planning procedures, targets action, is selective, and involves stakeholders, shifting attention to the process rather than the plan, and the ability to identify strategic issues, strategies, and actions through ongoing interaction. Through selectivity, involvement of the actors from the beginning of the process, and orientation to action, it seeks to overcome the presumed independence and omnipotence of the planner with respect to the context.

Albrechts (1999) assessed his direct experience of the Flanders Strategic Plan and indicated the 4 main tracks to be simultaneously followed in order to achieve successful strategic planning:

- vision construction
- stakeholder involvement
- the initiation of actions that offer an immediate sense of transformation

- raising public awareness.

Many years earlier, Lindblom (1975), in a little-known but very important essay, addressed the profound difference between strategic planning and what he defines as "conventional" planning, the latter still being considered the only form of planning in current practice.

In conventional planning, there is full confidence in the ability of the human intellect to understand and govern all phenomena; in strategic planning, planners are aware of their limits, which leads them to selectively assign their limited ability to govern the transformation processes, with a fair balance between what can be understood analytically and what can be understood through interaction. The title of the essay is in fact "The sociology of planning: thought and social interaction".

Lindblom introduces the definition of strategic planning with a series of examples.

To effectively plan the evacuation of a ship in the event of a fire, conventional planning operates by establishing meeting points, escape routes, and access to lifeboats, and by advising passengers on every trip regarding correct behaviour in case of emergency. But fire can occur in different parts of the ship and passengers who panic in the event of a fire do not remember or follow the directions. Strategic planners would still begin with the lifeboats, but instead of predicting improbably rational behaviours, it would work on the relationship between on-board personnel and passengers, instructing the former in managing panic situations and different potential locations of a fire. "They do not directly plan the solution to a problem but instead they plan the development of a capacity to invent a solution tailored to the particular form in which the problem eventually appears." (Lindblom 1975, p. 42).

Other examples include planning the war economy, control of atmospheric pollution, and planning the expansion of a highway system over 20-year period. In all situations, strategic planning seeks not to control everything, but to intervene with direct or indirect tools to trigger generative and learning processes rather than to plan a final solution.

Strategic planning therefore seeks first to understand the processes already in progress that are moving in the desired direction, rather than seeking to redesign entire systems; the goal is to harness the intelligence with which individuals and groups pursue their own objectives, and to induce changes that push actors towards improvement rather than replacing them entirely. This can be referred to simply as capacity building.

In the early debate on conventional versus strategic forms of planning, one must recall Albert Hirschman (1967), who in "Development Projects Observed" on a series of projects financed by the World Bank, proposed the "principle of the hiding hand": the aspiration to control all the variables in a complex project can prevent one from recognising that unexpected and unforeseeable events can occur in any situation and could cause a project to fail; but just as we fail to foresee these difficulties, we also fail to foresee our ability to react to difficulties once they arise. In other words, the hand that hides the difficulties we will encounter fortunately also hides our unexpected ability to react to them. If we knew all the difficulties that we are destined to encounter, we would not even attempt complex projects, and would thus also lose extraordinary opportunities that, in making the attempt, we are instead able to seize.

Hirschman is also the theorist of "possibilism", a practice-oriented position whereby in even the most difficult situation, one should seek latent resources and existing mechanisms that can resolve difficult situations, working on the possible rather than the probable. In "A Bias for Hope" (Hirschman 1971), he discusses how the trap of pre-established solutions and standardized intervention methodologies constitute an obstacle to research in every situation of unprecedented ways out.

Lanzara (1993), from a different disciplinary field, refers to "the negative capacity" as the ability to accept and live in uncertainty and disorientation, and to seize the potential and the possibilities for action that can be revealed at the moment of disaster.

2.3 Planning and Resilience

What does all this tell us about resilience and the forms of instability that are manifesting in society, the environment, and politics?

That our ability to organize reactions to more frequent shocks is increasingly necessary, but perhaps also that we must be able to assign our ability to plan with discretion.

There is an increasing need to collectively organize reactions, to plan responses, to prevent where possible, and to reconstruct in such a way that the fragile situations that led to the upheaval caused by a given trauma are not reproduced. Above all, there is a need to be ready for the totally unexpected.

All these different types of activities require different forms of action and planning. Different approaches depending on the characteristics of a given situation.

Here, however, it is worth distinguishing among the various problematic situations in which we can speak of resilience.

The distinction made by Donald Rumsfeld, then United States Secretary of Defense, on the different forms of uncertainty that characterize the problems that arise in crisis situations has become quite popular, also in social research:

> As we know, there are known knowns; there are things we know we know. We also know there are known unknowns; that is to say we know there are some things we do not know. But there are also unknown unknowns—the ones we don't know we don't know. And if one looks throughout the history of our country and other free countries, it is the latter category that tend to be the difficult ones.

It seems to me that this distinction helps us to organize our reasoning about the relationship between planning and resilience.

In the first place, therefore, there are things we know we know.

There are some problems related to recurring emergencies that can be addressed with serious forms of conventional planning. This is demonstrated by the development of Civil Protection in Italy, which has very quickly organized the emergency response to earthquakes, floods, and other natural disasters.

This requires good organizational capacity and a hierarchical and centralized command structure. It entails situations in which, following the reasoning proposed by Karen Christensen, there are no problems of consensus on the objectives and the intervention technologies have been refined over time: there is an immediately operational central command and an efficient network of territorial terminals, made up of public officials and volunteers with adequate equipment. In Christensen's scheme, these are situations where planning can be viewed as standardized programming, where the logic of technical rationality is dominant.

We know, however, that the moment of reaction to the emergency is only one phase, albeit a fundamental one in treating the crisis caused by a disaster, and we also know that it works mainly with reference to known types of disaster.

Then there are things we know we don't know.

We should recognize that, even with known disasters, when it comes to reconstruction and prevention, we still do not know how to intervene.

These are situations where the logic of standardized programming and technical rationality show all their weakness: post-event reconstruction is exposed to severe uncertainty. Political, social, and territorial dimensions come into play that are difficult to address with that approach.

During the post-earthquake reconstruction in Italy, there were conflicts of competency between institutions at different levels, opposing visions on how to proceed, and problems related to the availability of resources, so much so that in the case of most recent Italian earthquakes, the emergency structures designed to accommodate the displaced population for short periods have become permanent. If post-disaster reconstruction involving multiple actors and situations (for example, the 2016–2017 earthquake in Central Italy) is approached with a logic of conventional, centralistic, and hierarchical planning, not only can the recovery time be significantly extended for the classic problems highlighted in implementation research (Pressman and Wildavsky 1973), the approach may even block rather than facilitate interventions. The logic of the commissarial structure does not seem to be the most effective. If the sole decision maker is not equipped with an adequate structure possessing knowledge of the territory concerned, including its risk levels and organizational and social practices, they often become a bottleneck in the complex decision-making processes that preside over reconstruction.

Here, strategic planning à la Lindblom is helpful: experimental and bargaining approaches can follow concise guidelines and apply them differently according to specific territorial situations. Work on capacity building rather than directly planning the solution. Understand what virtuous processes are already in place that can be supported and generalized. Do not pretend to understand everything analytically, to govern through ordinances, but rely on interaction as a tool to discover potential and possibilities and to define effective reconstruction paths.

Even in the prevention of known risks, conventional forms of planning have encountered similar problems, which are aggravated and rendered unstable by a dwindling of the public attention that had brought prevention to the national political agenda in the immediate aftermath of a disaster. A perfect example from Italy is the Casa Italia program, activated in the aftermath of the 2016–2017 earthquake in Central Italy. The special task force, set up under the Prime Minister's Office, prepared a report (2017) aimed at launching a vast prevention program. As a consequence of the indications contained therein, a special Department was set up within the Prime Minister's Office to implement the program. Its operations, however, were the victim of numerous changes of government until it became so far removed from its original purpose that the issue slowly fell off the agenda of the country's priorities. Today, even the Department's mission has changed; originally a structure to promote prevention, its official website now announces that the aim of the Casa Italia Department is to support reconstruction.

All this is not surprising; this application of a form of conventional planning to subjects who are exposed to the instability of politics and bureaucracy was inevitably doomed to come up against what Donald Schön calls the dynamic conservatism of the very structures meant to preside over the implementation of innovative programs (Schön 1971).

This prevention program, inspired by the logic of strategic planning, could have been more effective if, as Louis Albrechts suggests, it had been simultaneously pursued along the four aforementioned lines: vision construction, immediate actions, involvement of the most important stakeholders, and reaching out for public opinion, so that public attention can be kept high.

Lastly, there are things we don't know we don't know.

The most evident case is the COVID-19 pandemic that we are currently experiencing, with the consequent confinement, illness, and death.

The calamity affects everyone closely and has found most countries unprepared; this demonstrates that the emergencies we may encounter in a context that has become permanently unstable are increasingly unknown and can take on unexpected and unpredictable forms.

What has been achieved in Italy by the Civil Protection procedures implemented by the government to face this terrible emergency, which has none of the characteristics of those of the recent past, from earthquakes to floods? I seem to understand very little, other than the identification of the head Commissioner for Civil Protection, who is faced with a completely new situation for which none of the good procedures developed over time by his organization have been decisive. Initially, the swift spread of the pandemic overwhelmed the government, the commissioner, the regions, and all the health structures. The Chief of Civil Protection was soon joined by other commissioners, and an exorbitant number of task forces were formed to deal with the explosion of a dramatic and unprecedented phenomenon.

This is only one example, but it demonstrates how much we need to build a response capacity for completely unknown situations.

The current pandemic, together with the other "unprecedented" situations we have experienced all over the world, from the devastating fires of Australia and U.S. in recent months, to the 2010 Tsunami and nuclear accident in Japan, to terrorist attacks in France in 2015, demonstrate that if conventional forms of planning are not effective, and we cannot even draw much from the indications of strategic planning, we still need to be prepared; we still need to develop a capacity of reaction to the unexpected.

Giovan Francesco Lanzara helps us by defining "negative capacity" as an activity that does not immediately seek solutions but lives in uncertainty and disorientation in order to seek the energies that can lead to a reaction. Albert Hirschman suggests taking a possibilist attitude that involves seeking latent resources that can be freed in order to get out of a situation that is apparently devoid of easy ways out. These are interesting positions that can offer inspiration, but it is not enough.

Andrew Lakoff, in a series of works (2007, 2017) talks about "preparedness", an approach to the imponderability of the calamities that can occur that aims not to avoid them, which is impossible, but to construct a reaction capacity that is valid in a variety of catastrophe situations. A form of planning that takes on the objective of preparing for the unexpected through scenario construction, the protection of critical communication infrastructure, the provision of devices that facilitate coping in different types of emergency, commissioning immediately activated alarm systems, designing systems for coordinating different subjects, and periodic verification of their operation.

In the context of the present pandemic, this set of actions would have been very helpful, from the provision of protective devices, to a clear outline of the relationships between subjects with competing skills, the functionality of alarm systems, and safeguarding and maintaining health infrastructure in the weak areas that have seen the most dramatic effects.

Preparedness for the incalculable disasters arising amidst growing social, economic, and environmental instability means constructing not the solution, but the capacity to react in the face of the things we don't know we don't know.

I therefore believe, in conclusion, that the resilience of societies and territories must continue to rely on planning in order to organize the response to shocks caused by different types of disasters. However, that planning will take on very different forms, depending on the nature of the problems we face.

References

Albrechts L (1999) Planners as catalysts and initiators of change: the new structure plan for Flanders. Euro Plann Stud 7: 587–603. https://doi.org/10.1080/09654319908720540

Albrechts L, Healey P, Kunzmann KR (2003) Strategic spatial planning and regional governance in Europe. J Am Plann Assoc 69(2):113–129. https://doi.org/10.1080/01944360308976301

Bryson JM, Roering WD (1987) Applying private-sector strategic planning in the public sector. APA J 53(1):9–22. https://doi.org/10.1080/01944368708976631

Burdett R, Sudjic D (eds) (2007) The endless city: the urban age project. Phaidon, London

Ceccarelli P (1978) La crisi del governo urbano. Marsilio, Padova

Christensen KS (1985) Coping with uncertainty in planning. J Am Plan Assoc 1, Winter, 63–73

De Rossi A (ed) (2018) Riabitare l'Italia. Le aree interne tra abbandoni e riconquiste. Donzelli, Roma

Florida R (2008) Who's Your City. Basic Books, New York

Hall P, Pain K (2006) The polycentric metropolis. Learning from mega-city regions in Europe. Earthscan, London

Hirschman A (1967) Development projects observed. The Brookings Institute Press, Washington

Hirschman A (1971) A bias for hope: essays on development and Latin America. Yale University Press, New Haven
Indovina F (ed) (1990) La città diffusa, "Quaderno Daest" n. 1. IUAV, Venezia
IPCC (1990) Climate change. The IPCC Response Strategy, United Nations
Lanzara GF (1993) Capacità negativa. Competenza progettuale e modelli di intervento nelle organizzazioni. Il Mulino, Bologna
Lakoff A (2007) Preparing for the next emergency. Public Culture 19(2):247–271. https://doi.org/10.1215/08992363-2006-035
Lakoff A (2017) Unprepared: global health in a time of emergency. University of California Press, Berkeley
Latour B (2018) Down to earth. Politics in the new climatic regime. Polity Press, Cambridge UK
Lindblom CE (1975) The sociology of planning: thought and social interaction. In: Bornstein M (ed), Economic planning East and West, Cambridge Mass, Ballinger
Struttura di Missione Casa Italia (2017) Rapporto sulla Promozione della sicurezza dai Rischi naturali del Patrimonio abitativo, Presidenza del Consiglio dei Ministri. https://www.governo.it/sites/governo.it/files/Casa_Italia_RAPPORTO.pdf
Pressman J, Wildavsky A (1973) Implementation. University of California Press, Berkeley
Rittel HW, Webber MM (1973) Dilemmas in a general theory of planning. Policy Sci 4(2):155–169. https://doi.org/10.1007/BF01405730
Rodríguez-Pose A (2018) The revenge of the places that don't matter (and what to do about it). Cambridge J Reg Econ Soc 11:189–209. https://doi.org/10.1093/cjres/rsx024
Schön DA (1971) Beyond the Stable State. Norton, New York
Schön DA (1983) The reflective practitioner. Basic Books, New York

Chapter 3
Risk Mitigation and Resilience of Human Settlement

Scira Menoni

3.1 Risk Mitigation and Resilience: A Theoretical Perspective

Risk as a construct is recent in our culture. In his marvelous book, Bernstein (1998) told "the remarkable story" of this concept, that has evolved over centuries until the XIX, when mathematical formulas were applied to the disciplines of statistics and to the theory of probability and probabilities were used to assess the chances of success or failure of action under uncertain conditions. The use of probabilities to measure risk is therefore rather new, despite of the fact that ancient Greeks already disposed of the mathematics it requires, at least at the basic level. This means that risk, intended as the probability of damage consequent to an event or an activity is relatively novel and only recently associated to engineering practices of developing daring structures and artefacts with the potential of producing harmful side effects when stressed by natural extremes or because of the dangerous processes entailed. Risk is an anthropocentric concept, as it addresses damage to people and to assets that are at the same time located in a dangerous area, exposed to a given natural or man made threat and vulnerable to the latter. Methods for identifying and measuring hazards, exposure, and vulnerability have improved in the last decades, but there are still significant challenges in such assessments when applied to complex systems as cities are. In particular, the concept of vulnerability has emerged only lately as a crucial factor depicting the capacity to face and sustain stress by a given system or system's component as independent from damage and exposure. For damage to occur vulnerable exposed systems must be affected by a hazardous occurrence. Vulnerability differs from exposure as the former points at weaknesses, fragilities,

S. Menoni (✉)
Department of Architecture, Built Environment and Construction Engineering, Politecnico di Milano, Milano, Italy
e-mail: scira.menoni@polimi.it

© The Author(s), under exclusive license to Springer Nature Switzerland AG 2020
A. Balducci et al. (eds.), *Risk and Resilience*, SpringerBriefs in Applied
Sciences and Technology, https://doi.org/10.1007/978-3-030-56067-6_3

proneness to damage whilst exposure refers to the numerical quantification of people and assets or the appraisal of economic values located in a hazardous area.

Different types of vulnerabilities have been analysed in human settlements, including physical vulnerability of artefacts and infrastructures, social and economic vulnerabilities. With the term systemic vulnerability the fragilities stemming from the complex interaction of interdependent systems are pointed at, as such interconnection is larger in cities and metropolitan areas where assets, people and functions are concentrated.

It has been suggested that the concept of vulnerability is linked to an engineering understanding of risk. However, this view can be easily counteracted as a very large body of research has been devoted to understanding the vulnerabilities of social groups, individuals, organisations from a variety of perspectives and significant advancement have been achieved in appraising how different economic activities, sectors, combination of the latter, may be more or less prone to suffer losses in the aftermath of disasters (Schneiderbauer et al. 2017).

3.1.1 The Resilience Concept: A Framework

There is no doubt though that partially overlapping with the evolution of the risk concept and the progressing investigation of all the variable conducive to damage, and partially deriving from other streams of research, including climate change adaptation, growing attention has been devoted to resilience as a term partially independent and partially overlapping with vulnerability and risk.

Similar to vulnerability, though, also resilience has multiple meanings and has opened the floor to very wide discussions on semantic aspects. Here we propose a framework acknowledging the various facets of resilience as shaped by different disciplinary perspectives. The aim however is not providing another conceptual interpretation, but rather to identify key aspects that can be useful to operationalize the term when applying it to human settlements in order to reduce damage and losses to communities and assets consequent in particular to stresses of natural origin.

Table 3.1 summarizes the most relevant disciplines that have provided interpretations of resilience in a figurative way, alluding at different degree to its root meaning in mechanics being the energy absorbed by an elastic material under stress. The closer to this root meaning is the understanding of resilience as the capacity to bounce back after a severe shock as the one provoked in a system by the impact of a natural extreme. For example, Bruneau et al. (2003) analyzing the resilience of critical infrastructures following earthquake, measure it as the time needed to recover their full functionality. Time for recovering has become a standard in similar approaches to resilience. Such interpretation is rather different from that proposed by Hollnagel et al. (2006) or by Park et al. (2013) despite of having been developed by engineers as well. In fact, the resilience engineering approach suggests that complex systems are exposed not only to known risks, for which traditional probabilistic approaches are viable, but also to surprises and unexpected occurrences that strain

Table 3.1 Selected theoretical approaches to resilience

	Scientific and technical domain	Psychology and psychiatry	Echology and climate change studies	Systems engineering Resilience engineering
Proposed interpretation	[a]Resilience is mainly studied with respect to the performance of critical infrastructures under stress. Resilience is measured in terms of time needed to recover after a severe stress	[a]Resilience is related to post traumatic rehabilitation. Psychologists and psychiatrists study how people recover from disasters and other traumas and search for conditions that favour resilience	[a]Ecologists define resilience as the capacity of ecosystems to sustain stress without losing their functions and find a new dynamic equilibrium	[a]According to the resilience engineering approach it is important to study how complex systems react to unexpected stress it has not been designed for. It accepts the possibility of failure, but aims at failing gracefully, without catastrophic cascading effects
Selected key references	Bruneau et al. A framework to quantitatively assess and enhance the seismic resilience of communities, in "Earthquake Spectra", vol. 19:4 (2003)	Norris et al., "Community resilience as a metaphor, theory, set of capacities, and strategy for disaster readiness". American Journal of Community Psychology 41(1–2): 127–150 (2008)	Holling C., Resilience and stability of ecological systems, Annual Review of Ecology and Systematics, vol. 4 (1973)	Park et al., Integrating risk and resilience approaches to catastrophe managementin engineering systems, in Risk Analysis, vol. 33, 3 (2013)
		Paton, D. Community Resilience: Integrating Individual, Community and Societal Perspective (2008)	Gunderson L., C. Holling, Panarchy. Understanding transformation in human and natural systems Island press (2002)	Hollnagel, E., D. Woods, and N. Leveson (eds) (2006). Resilience Engineering: Concepts and Precepts. Aldershot, UK: Ashgate Publishing Limited

systems in a way that requires flexibility and unprecedented response to overcome the shock provoked by a disaster. Resilience is that capacity of complex systems to adapt to changes, to manage uncertainty and to bounce forwards towards new configurations. Resilience lies at the interface between technical and social systems, therefore resilience engineering requires the understanding of how physical aspects are interlaced with cognitive, informational and cultural resources. Differently from traditional approaches, resilience engineering does not demand from system not to fail but rather "to fail gracefully", avoiding catastrophic cascading effects. The most widely known and quoted interpretation of resilience is perhaps that developed in the ecological domain, that shares many similarities with the one adopted by experts of climate change, that follows the classic book "Panarchy" by Gunderson and Holling (2002). Stemming from applications to ecosystems, this interpretation has been extended to human and social systems by the same authors, indicating the ability of systems to be stressed without losing their functions and instead of being disrupted reaching a new state of equilibrium. This interpretation has been adopted by many geographers, sociologists of disaster studies, planners, because whilst avoiding mechanistic understandings of resilience as the ability to bounce back to a state that in many cases was neither desirable nor sustainable, it is wide enough to be adapted to many conditions of risk.

Less acknowledged, albeit in our view fundamental in the case of disasters that entail large human losses, is the interpretation developed within the disciplinary domains of psychology and psychiatry. In the latter, resilience is the capacity of individuals and communities to overcome traumas they have endured in a way that transforms the negative experience and sometimes irreversible wounds into a positive reaction, consisting for example in devoting time and efforts to volunteering activities or choosing a profession devoted to save and protect lives. In fact, disasters certainly represent a collective trauma, especially when deaths occurred. The deeper the trauma the less bouncing back is possible or desirable.

One may legitimately ask why was there the need to introduce resilience as vulnerability is already addressing many components of the societal response to disasters. In fact, we can still consider as valid the finding of Cutter et al. (2008) based on the analysis of a vast literature, according to which the relationship between vulnerability, coping and adaptive capacity, resilience neither is clear nor straightforward. Within the EU funded project Ensure we have favoured one of the interpretations included in the vast array of possibilities listed by Cutter et al. (2008). According to the latter, resilience and vulnerability can be only partly considered as opposite terms and in particular the resilience of critical infrastructures can be considered as the opposite of systemic vulnerability. However, the two terms are also independent and complementary to each other, meaning that a system can be both vulnerable and resilient at the same time. According to the Ensure project interpretation, vulnerability refers more to the short term response, both in terms of physical resistance of assets and people to a stress, whilst resilience addresses the longer term recovery and reconstruction capacity. Resilience relies on resources, capacities, ability to learn from the experience lived through in a disaster that is projected into the future, into the new configuration that cities and communities will have after reconstruction. In this regard the proposed definition provided by Norris et al. (2008) drawing on psychology and psychiatry literature is enlightening "community resilience [emerges] from a set of networked adaptive capacities" made of community competence, social capital, economic development, information and communication. The Authors specify that the networked capacities rest not only on the interlinkages among the main components listed above, but also on the resources attributable to each of those components, such as flexibility, creativity, availability of participatory approaches, trustfulness, etc.

Despite of the differences between the definitions and understandings provided by the different disciplines as in Table 3.1, there are still significant commonalities. In particular, most approaches consider the dynamicity of resilience, that is not considered as a stable feature but rather as a process that can be favoured by some conditions and strategies. In fact, resilience can be attributed both to the system and to the process, as resilience is not a property of a system, but rather the way in which it behaves.[1] Resilience requires a positive reaction to a negative experience and relies on available knowledge and information as vital resources and on the capacity to learn lessons and adapt to new configurations and dynamics. As in the framework by Norris et al. (2008), though, social and human capital will not suffice; financial resources and economic factors are essential for a resilient response to a disaster.

One important aspect at the base of resilience approaches, especially those grounded in psychiatry and engineering, regards the possibility to identify factors that not only characterize a resilient behavior but can foster it. On those premises for example the UK government decided to create the new Cabinet Office UK Resilience in the aftermath of the September 11 terrorist attack. But as discussed by Medd and Marvin (2005) whilst "the governance of preparedness is based less on responding to an emergency and more about a strategy of 'building resilience'" it was not that easy to operationalize such strategy and translate a political mandate into action (see also Coppola et al. in this volume).

Whoever wishes to use the resilience approach to practically enhance the capacities of a system, in our case human settlements, needs first to develop methods and procedures to represent and possibly measure resilient features and behaviours and second to suggest which of the latter can be actually triggered and/or encouraged by ad hoc strategies and arrangements.

3.2   Mainstreaming Knowledge on Risks and Mitigation into Land Use and Urban Planning

In 2001 three eminent scholars (White et al. 2001) who have shaped the history of modern disaster studies wrote an article with a rather worrying title "Knowing better and losing even more: the use of knowledge in hazard management". Among the

---

[1]From an interview to Olav Grotan, researcher at Sintef in the context of the Know4drr project (www.know4drrproject.polimi.it).

reasons they were analyzing for this apparently contradictory situation, poor land use planning was considered key. According to White et al. "many programs have been developed over the years to manage natural hazards from flood control to earthquake building codes, to land use regulation, to insurance. Despite the many good intentions these programs often seem to have failed to produce the anticipated results. Flood control projects have served to encourage more floodplain development. Federal disaster assistance in some areas has encouraged continued occupancy of areas with repetitive losses. Land use regulations have been opposed through aggressive legal action and have often been applied with a lack of conviction, or have been subject to frequent variances obtained by political means" (p. 90).

Despite of the fact that many international organisations[2] have highlighted the importance of land use and urban planning as a key non-structural mitigation measures with a very large potential of diminishing overtime all the components of the risk function (Melchiorri 2016), still planners have been scarcely interested on the matter for quite a long time and prevention has not become an integral part of ordinary plans and projects. In the last years planners have invested more into the issue, especially through the mediation of the resilience concept that appears to be more appealing than risk prevention and via policies and strategies addressing climate change adaptation. Hopefully in the next future the two streams of disaster risk reduction and climate change adaptation will address overlapping and complementary aspects easier to integrate into both strategic and everyday decisions regarding urban and regional development. Seeds of such needed integration can be found in recent project calls of the European Commission, in some international conferences (Galderisi and Colucci 2019), in analyses of current practices demonstrating the fallacies consequent to the separation of policies related to climate change adaptation on the one hand and disaster risk reduction on the other especially at the city level (Storbjörk 2010).

A fundamental step in the direction of triggering a change in current practices in planning requires the revision of educational and training programs, introducing in a more decisive way topics related to the analysis and management of disaster and climate related risks. One of the main obstacles in both training programs in policy sciences and in architecture stems from the difficulties in embedding and blending in the same courses technical aspects that are fundamental to understand exposure, hazards, vulnerability and the analysis of the socio-political drivers of risk and lack of resilience. It must be also acknowledged that this represents an obstacle that is encountered in all modern challenges that would require a truly interdisciplinary and multi-stakeholders approach, that, despite the many claims, still remains a difficult and rarely accomplished endeavor.

Whilst part of the responsibility for the poor consideration of risk components in ordinary planning lies with planners, there are also other compelling reasons for it. In the Armonia project funded by the EU Commission under the VI FP in particular, some tasks were devoted to propose standardized legends of hazards technical

---

[2]Ranging from the Un Habitat Agenda to the Eclac and UNECE Guidelines for spatial planning to the UNDP call for integrating land use planning and sustainable development.

maps usable by planners. By usable it was meant at the scale at which plans are prepared, distinguishing between regional and local scales (Margottini and Menoni 2018) and focusing on aspects of geology and geomorphology that are of immediate relevance to planners (avoiding details that are meaningful for geologists but are not really essential to understand the specific problem at stake in a given zone). In the latter two decades significant steps forward have been made in the direction of mainstreaming disaster knowledge and information on different aspects of risk into the typical analyses conducted by planners to describe and characterize settlements they are preparing plans and projects for.

3.2.1 Measuring Risk and Its Variables in Cities and Human Settlements

In order to better understand how risk variables can be measured to be of use in cities and settlements management some examples will be provided in the case of floods. Parameters such as peak discharge, water depth and velocity, presence of sediments and contaminants, permit to characterize the hazard; exposure refers to the quantitative measure of number of people (inhabitants, workers, visitors) and goods (as well as their economic value) that are located within the floodplain area; vulnerability refers to the assets located in the first floors of buildings and to people living in basements. The human population is exposed to the threat, even though at different degrees given the concentration and density in given places and areas (typically large metropolitan areas and within the latter transportation hubs, services and facilities hosting a large number of people temporarily or permanently). Vulnerability can be referred to either specific characteristics that make some places more vulnerable than others (for example low passages under a highway or a railway where the larger number of victims occur during a flood) and the intrinsic susceptibility of certain social groups (by age, mobility conditions). It may be reminded the case of the camping in Soverato Southern Italy in the year 2000 where thirteen disabled people died trapped by a flash flood that occurred in a seasonal river. In the natural hazard domain, disaster risk reduction measures may target all of the three main components of risk, distinguishing between structural and non structural ones. The former generally apply to the hazard component, with the aim to reduce the severity/frequency of the hazard, in the case of flood by constructing levees, containment basins, dams. Some structural measures aim at reducing the vulnerability of buildings, for example elevating above the expected water depth. Non structural mitigation measures address exposure, in the attempt to limit the number of people/goods in a hazardous area (see the interesting example provided by Curci in this volume), for example by relocation, and vulnerability, acting on land use and spatial planning through a careful combination of zoning, allowed functions and type of assets combined with emergency planning and preparedness. The basic assumption is that in any case, no matter what mitigation measure is put in place, zero risk cannot be achieved. In the "disaster science" domain, scholars identify different types of damage, that often are enchained to each other (Menoni et al. 2017). Direct physical damage is the one provoked by the hazard on vulnerable exposed systems. In the case of floods, it is for example the death toll, the number and value of infrastructures, homes and economic activities including agriculture that are flooded or contaminated by incoming water. Indirect damage is related instead to second order consequences that are due to systemic interconnections among systems and systems' components that amplify the consequence of physical damage, even in cases where the latter has occurred only in a small part of the system. Indirect damage includes for example business interruption, social discomfort and psychological impact due to the trauma of lost relatives and friends, etc. The indirect damage is generally associated with the event and the physical damage, but sometimes it is the consequence of mitigation measures that are put in place. In the case of floods, retainment basins that are constructed upstream to protect cities downstream are clearly infringing on land that may be devoted to other uses thus harming communities and/or owners of the land that will be occupied by the protection structures. Even in cases when such retainment basins will be used for agricultural production, in case of flood, damage will be suffered by farmers. Even in case the so called nature based solutions are selected, someone will be still paying the costs for them. In an enlightening book chapter, Warner et al. (2012) discussed the case of The Netherlands, showing that the well known and advocated policy "space for the river" has encountered fierce opposition by those who had to pay for not only giving up potentially profitable uses but in some cases obliged to move to another area.

In the last years, significant research and practitioners' effort has been invested into developing methods to assess the various components of risk. One of the most advanced tools is the French Risk Prevention Plan that at the local, municipal scale, characterizes risk by identifying and assessing the hazards threatening a given municipality, the exposed assets with some attributes of vulnerability. Following the seismic swarm that affected Christchurch in the years 2010–2011, increased interest has been devoted to land use and spatial planning in New Zealand, a country that had already in place several initiatives on prevention and preparedness.[3] Saunders and Kilvington (2016) developed a rather sophisticated process entailing the interaction with relevant stakeholders to appraise together with them the implications of the presence of multiple hazards in a given territory and the potential impacts they may cause on differently vulnerable assets and infrastructures. A limitation that still persists relates to the relatively poor or too rarely considered use of comprehensive scenarios as a really informative tool for planners. In the latter, instances of potential impacts in an area are provided, combining risk factors, different inputs in terms of hazard severity and frequency and decisions made to reduce exposure and vulnerability. There are some shortcomings in the current way scenarios are used (following Curci in this

---

[3]See for example the program funding research on natural hazards mitigation: https://www.civ ildefence.govt.nz/cdem-sector/cdem-research-/natural-hazards-research-platform-/; and Resilient Organisations https://www.resorgs.org.nz/about-resorgs/; the initiative on lifelines resilience at different levels https://www.nzlifelines.org.nz/.

volume). First, scenarios are often developed having in mind emergency planning as a sole field of application, second low-probability events are perceived as being less urgent and therefore action is delayed and, last but not least, there is limited capacity to embed knowledge available from past experience. Elsewhere (Mejri et al. 2017) we have pointed at the fact that planners are often trained to design and make projections in a landscape that is perceived as rather stable, neglecting the fact that some natural hazards may dramatically and irreversibly change it in a very short time duration. The erosion of beaches following hurricanes can be mentioned here or denuded slopes after tephra deposit.

Many aspects remain unsolved, the most important of which is probably the lack of coherent and agreed upon methodologies to address the multiscalar aspect that is fundamental for planning that needs to act at different albeit connected spatial scales with provisions that are coherent with each other and addressing different factors at the same time. For example, at the large regional scales, provisions for critical infrastructures and lifelines must be decided; at the local scale, physical vulnerability concerns are key as artefacts are actually built in a specific place and with specific materials and techniques. At regional scale, considerations of distribution and weights of exposed people and assets, and of the qualitative layout of settlements in terms of accessibility, shape and neighborhoods pattern are important to define levels of systemic vulnerability, intended as the propensity to indirect, systemic damage. The latter is due to the interdependency and lack of redundancy of for example transportation or power networks that serve a specific area.

3.2.2 Resilience Applied to Human Settlements

Referring to the framework provided in Fig. 3.1. it can be seen how the different interpretations of resilience have actually been all applied and further developed to human settlements. The interpretation of resilience as capacity to bounce back and return to normalcy has been mainly applied to urban critical infrastructures and has considered the interaction between vulnerable constructions and lifelines (Cimellaro 2016; Cimellaro et al. 2019). The understanding of resilience as a collective healing process has been certainly embraced by Vale and Campanella (2005) in what can be considered already a classic book on urban resilience. By examining the reconstruction of cities following wars, fires, earthquakes, they showed rather convincingly that human centred recovery is essential, often symbolized by monuments and memorials representing an essential step towards the resilient overcoming of a tragedy. The theory proposed by Gunderson and Holling (2002) gave rise to many subsequent streams and has been perhaps the most referred to in studies and researches regarding urban and cities resilience. Their theory proved to be particularly appealing as it addresses the dynamic non linear evolution of human settlements that has undergone a dramatic impulse in the more recent years with more than half of the human population now living in cities. It also provides an explanatory framework for what has been pointed at as evolutionary approach to resilience (Davoudi et al. 2012)

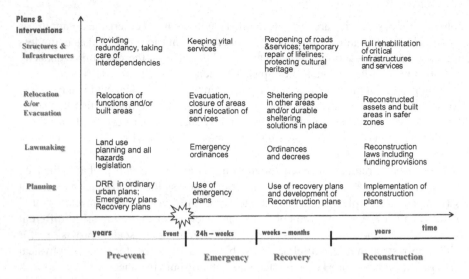

Fig. 3.1 Plans and activities across the disaster cycle

capable of embedding the analysis of the interaction between technical and social systems that occurs at different levels (Wang and Yamashita 2015). Planners have instead payed less attention to the resilience engineering approach, despite of the fact that it provides relevant insight into the complexity of current dynamics and features of human settlements. In a recently concluded EU funded project, Educen, we have provided such link referring in particular to the work of Park et al. (2013). "Cities share with complex systems virtually all features that characterize them as such. First, given the nonlinear interrelationships between systems and components, it is impossible to predetermine how the latter will interact with each other in the future or under changed conditions. Second, there is the 'path-dependency' of decisions that are made, particularly those regarding locational issues and the establishment of new urban functions in a previously not urbanized area. Third, the difficulty to forecast the response to external forcing such as that imposed by natural hazards or manmade severe accidents. Other important aspects that need to be considered are: the fast evolution of the 'urban' particularly in the more recent decades, implying differential development dynamics; the multilayered governance of cities, resulting from a deeply transformed geography. In fact, in the new urban landscape fringes tend to blur into rural areas making it very hard to define a clear cut border between what is 'city' and what is not" (Menoni and Atun 2017).

Despite of the large success of the concept of resilience in recent literature on urban and spatial planning, operationalizing the term and finding indicators and criteria to assess its presence or absence in a given city not to mention practical tools and methods to put it into practice proved to be rather hard. Also in this case a significant gap originated between a practitioners' approach that has been adopted by public administrations in a rather pragmatic fashion and the well elaborated but not applied theories developed by researchers in the field. Since the 2000 decade in fact, some initiatives have focused on making cities more resilient, whilst less attention has been devoted to rural areas, or to areas that are not central, such as mountain or island communities or towns that remain peripheral to the large commercial and financial networks. Among such initiatives, the 100 Resilient Cities sponsored by the Rockefeller Foundation has nudged a method developed by Arup[4] consisting of a rather standard set of indicators that has the merit though to operationalize the term resilience, permitting to develop a structured assessment of cities and consequently define a strategy toward the improvement of those indicators that proved to lay on a poor performance threshold. The challenges implied in contextualizing such method to specific social, cultural and political settings are described for the cities of Milan and Rome in the contribution of Coppola et al. in this volume. A number of European funded projects have proposed methods and tools for assessing cities resilience. Results though are still of mixed quality, so that there are still open calls asking for more integrated approaches to assess not only whether or not cities and human settlements display resilient features but also provide guidance on how to advance in order, among other goals, to achieve a better integration between measures aimed at reducing direct and indirect damage to both risks and impacts of climate change.

3.2.3 The Need for a Truly Interdisciplinary and Multi-stakeholders Approach

What is probably still missing is a truly interdisciplinary and multi-stakeholder approach. It is a sort of "common sense" in the scientific community working on risks, hazards, prevention, that an interdisciplinary approach is required, and for a number of good reasons. Some are rather self-evident: the multiple competences needed to study different phenomena (sometimes enchained), the various components of risk (hazard, exposure, vulnerability) that call for a variety of experts. Other reasons are less banal: vulnerability and resilience of complex systems must be studied across multiple spatial and temporal scales. No single scientific community or expertise is able to address those issues satisfactorily. With respect to the past, it can be said that interdisciplinary research has been accomplished; several teams formed by researchers and practitioners with disciplinary backgrounds have worked together in projects. Nevertheless, obstacles and constraints are still experienced (see Nicolson et al. 2002 and Lélé and Noorgaard 2005) and the call for different definitions such as "transdisciplinary research" will not suffice to advance otherwise traditional practices. In fact, what the above quoted authors suggest is that there is the need to rethink interdisciplinary work as one in which not only solutions are searched collectively, but the same problem to be tackled must be framed together, jointing and blending the different perspectives that can enrich the overall understanding of the problem at stake. Ironically, though, such collective enrichment

---

[4]https://www.arup.com/perspectives/city-resilience-index.

come at the expense of the possibility for each expert to go in depth into his/her own discipline and this, coupled with how careers are conceived nowadays especially in academia, counteracts truly collective efforts. It is not by chance that innovation comes at the frontier of disciplines when the interdisciplinary endeavor really works, starting from the mutual respect and effort to contextualize one's knowledge in the knowledge of the other. This blending and combination of expertise is essential in tackling complex problems where monodisciplinary approaches are very likely to fail. According to McGlade and van den Hove (2013, p. 415): "it is unsurprising that planning and management institutions have been unable to respond to crises or change, as in many instances, the organizations are suffering from a chaotic mixture of information, analysis and interpretation with no paradigmatic structure in which to incorporate all the various forms of scientific, interdisciplinary, and indigenous knowledge." In a post-modern society, individuals are receiving information from an enormous number of sources. "In this way we can see a form of second-order science emerging in which individuals must rely on other peer groups and experts to be able to evaluate the information within their own domains of expertise."

As for knowledge that is necessary to mainstream risk mitigation and resilience into planning, two common difficulties are often experienced in the dialogue between experts in earth sciences, engineers and planners. One derives from the need in planning to combine technical and scientific information with more qualitative descriptions of social and cultural aspects of cities. Ginzburg (1980) suggests that a fundamental irreducible difference between what he calls Galilean and social sciences lies in the treatment of the individual as opposed to the typical, treatable in statistical (quantitative) terms. The possibility or not to treat statistically individual elements is also at the base of the capacity to predict the behavior of a variable, the evolution of a given phenomenon. The second difficulty relates to the scales at which different disciplines study their object of investigation. As suggested by Root and Schneider (1995) «the scale at which different research disciplines operate make multidisciplinary connection difficult and necessitate devising methods for bridging scale gaps» (p. 334).

3.3   Tools for Enhancing Resilience Through Land Use and Urban Planning

Operationalization of knowledge in risk and resilience is essential in order to provide solutions that are manageable and conducive to urban patterns and organization of cities' functions that are more resilient and less disaster prone. Planners have at their disposal a number of tools that they ordinarily use in order to develop better urban expansion, provide adequate level of services and infrastructures both in central and more peripheral areas, amenities that contribute to the well being of communities, both residents and visitors. In the following some of the tools that have been identified as key by scholars and practitioners in planning will be discussed as well as some relevant constraints to their effective and successful implementation for disaster risk reduction and resilience enhancement.

3.3.1 A Framework Comprising Tools for Land Use and Urban Planning in Hazardous Areas

Table 3.2 results from the combination of two frameworks that have been already proposed in the past. One derives from an original scheme developed by Bolton et al. (1986) where a number of instruments at the disposal of planners were illustrated comprising both purely urban planning and design tools with other that are of a more economic nature and we prefer to consider them as capable of advancing and favoring the implementation of plans and projects.

The second framework has been proposed by Menoni et al. (Menoni and Margottini 2011, chap. 4) displaying different structural and non structural mitigation measures and clearly pointing at the risk component such measures are actually addressing. More recently the same framework has been re-worked in Pesaro et al. (2018) with the ambition of identifying key factors and criteria for assessing both the effectiveness and the cost/benefit ratio of proposed measures with a specific focus on non structural mitigation.

The table is not exhaustive, it is representative of some of the tools that are more familiar to planners and better usable to prevent risk and reinforce resilience. The table is organized so as to first describe the tool in the second column, then explain how the measure is addressing hazard, exposure physical and systemic vulnerability factors respectively in the third, fourth and fifth column. The sixth column is highlighting how the tool can be conducive towards more resilient settlements and the last column briefly hints at both positive achievable aspects and limitations of the tools.

Zoning is certainly a rather wide-spread tool that exists in different forms in most countries. It permits to diversify areas based on their prevalent use distinguishing between a variable number and detailed typology of uses, very roughly categorized as residential, productive, commercial, services, open spaces, etc. Definition of acceptable uses is an important driver of risk or prevention, as all factors of risk are actually affected by such decisions. Through zoning one may even "create" new hazards, for example when a potentially unstable slope is expected to be urbanized and then subject to larger weight (buildings) or to cuts at the slope base when new roads are foreseen for guaranteeing connection with the newly built areas.

Subdivision standards are more fine tuned tools that can be used to control number and concentration of people in dangerous areas and to address different indicators of physical and systemic vulnerability.

Relocation is clearly a very costly measure that has been nevertheless considered by different legislations in many countries at least for the most hazardous installations located in areas prone to natural hazards and/or to built areas that are often affected

Table 3.2 Tools for reducing risks in and through resilience oriented plans

Tool	Description	Hazard factors	Exposure factors	Vulnerability physical and systemic	Resilience	Positive aspects and limitations
Zoning	Division of the area in sub-zones to accomodate different urban function and land uses, specifying norms for development in each one	New development can "create" some hazards (hydraulic or landslides)	Zoning can increase or decrease exposure in hazardous zones. Exposure can be calibrated based on a fine tuned hazard zoning acting on both surfaces and volumes standards	By deciding open spaces, location of transportation networks and nodes, systemic vulnerability can be addressed	Scenarios of optimal zoning in areas that are already urbanised can be identified for pre-event recovery plans	A multihazard approach can be taken in zoning new development but also in redeveloping some areas. It is a bit rigid tool to follow the dynamic changes of cities including changes in the use of buildings and spaces
Subdivision standards	Standards related to the specific features for the urban design of areas to be developed, redeveloped or restored		By defining acceptable volumes, occupied surfaces, height of and distance between buildings, subdivision standards are a key tool for addressing concentration of people and built up areas	By defining issues such as morphology, road network density, width, etc such standards can address significantly both physical and systemic vulnerability of urban areas	Subdivision standards can be revised after a disaster in an easier way and more accepted by owners	Subdivision standards infringe directly on property rights so there is the need to design them carefully and according to a participatory approach

(continued)

Table 3.2 (continued)

Tool	Description	Hazard factors	Exposure factors	Vulnerability physical and systemic	Resilience	Positive aspects and limitations
Relocation	Partial or total relocation of urban neighborhoods or entire settlements		Diminshes the exposure in a hazardous areas	Special care needs to be devoted to resettlement in the new areas	By avoiding or reducing exposure in hazardous areas it reduces the amount/severity of damage thus will require less efforts in recovery	A very effective disaster risk reduction measures, though very costly both economically and socially
Location of critical infrastructures and lifelines	New infrastructures and facilities are often public or partially public. Such choice needs to be taken strategically thinking not only of local concerns but also about the position of the city in the global networks and its economic role	Some lifelines can create hazards (roads in mountain slopes; gas conducts potentially provoking a natech)	Decision on location can be particularly critical not only for the infrastructures per se but also because it is strategic in attracting private investment and future	Should be built according to appropriate state of art codes, taking into consideration in the design multi-hazard concerns	In the aftermath of a disaster having a plan for better and safer critical infrastructures is key to reconstruct improving pre-event conditions, modernizing some networks and facilities	Monitoring is required overtime to assess dynamic conditions of risk in the area

(continued)

Table 3.2 (continued)

Tool	Description	Hazard factors	Exposure factors	Vulnerability physical and systemic	Resilience	Positive aspects and limitations
Defense infrastructures and structures	Structural mitigation measures such as levees, consolidation of landslides, avalanches defenses are aimed at reducing the hazard severity and/or frequency. Such infrastructures can be "nature based" and combine ecological and risk mitigation concerns	Structural mitigation measures such as the creation of retainment basins requires to empty areas that may be built up. Among such structural measures, make space for the river is an eminent example of nature based solutions	Some are aimed at reducing the physical vulnerability of artefacts, like building codes. Some measures such as green roof are considered "nature based" and combine relief from heat waves and rain water retainment		Any preventative measure if correctly designed can reduce the severity and extent of damage, thus permitting a faster and less costly recovery	Structural measures proved to be extremely effective over the history, they may provide a false sense of security and encourage development in highly hazardous zones. Further as most structures they require maintenance overtime

by floods or volcanic eruptions (in Italy for example it has been introduced following the Soverato incident mentioned above).

The provision of new infrastructures and services in safe areas is not only a good practice especially if public money was spent, but they also attract private investment towards safer areas. The weak point of this recommendation is that too often decisions regarding critical infrastructures, regarding for example large transportation nodes and networks, lay often outside the realm of ordinary planning instruments, especially at the local level. This way addressing both physical and systemic vulnerability at the local level is hampered by projects that are designed at much higher scale, too often with little or no participation of local scale planners and public administrators (Greiving et al. 2016).

Similar to large infrastructures also structural defenses such as levees or retainment basins are decided and designed outside ordinary plans. They could instead provide a crucial added value if integrated into such plans. On the one hand urban plans may help provide solutions that are better from a social perspective and as far as their integration into the landscape is concerned. In fact, the plan could work both as a coordinating tool of structural defenses better integrated in terms of design and land uses (including land uses in the vicinity of the structures) in the territory where they are constructed, and as a non-structural measure per se (acting on exposure, vulnerability and enhancing resilience with all the other tools). In the last years "nature based" structural measures have been promoted as an alternative or in combination with more traditional engineered ones, with the ability to combine ecological and risk reduction concerns.

The underlying assumption of the table is that all the tools can add resilience to the system, either if they are adopted in the recovery and reconstruction phase or if they are part of a strategy that considers different alternative scenarios that may occur and the ad hoc responses that can be foreseen before an incident.

The tools that have been briefly described are then utilized in different contexts and at different time scales of a disaster. Ideally we would like them all to be deployed before a disaster occurs. However, experience tells us that most will be effectively used in the aftermath of an event that will provide a window of opportunity to embed prevention and resilience concerns into recovery and reconstruction. Figure 3.1 shows the different use and adoption of plans and intervention across the so called disaster cycle. The figure builds on four main dimensions: the plans, the legislative apparatus that is put in place or is already established for each phase, activities and decisions of temporary or permanent relocation and provisions for critical infrastructures and structures (such as for example cultural heritage).

Not all interventions and plans are under the direct control or are part of the traditional toolbox of planners. For example, emergency plans may benefit from better coordination with urban plans and also from expertise aiming at better representing in maps various areas and strategic assets, but are outside the regulatory field of land use and urban planning.

Inclusion of preventative measures in plans as well as reconstruction visions and strategies are more clearly at the core of planners' activities and objectives. In more recent times, larger attention has been devoted to the phase of recovery, skewed between emergency and reconstruction for which generally no specific tools, plans or legislation has been prepared insofar. Yet in recent events, including the Christchurch and Central Italy seismic swarms that have occurred respectively in 2010–2011 and 2016–2017, several recent hurricanes in the USA and Caribbean Islands, floods in Europe, it has become clearer that specific provisions are needed for recovery, both in terms of ad hoc legislation, as well as the planning and design of the temporary city that originates form partial relocations and temporary sheltering solutions. The temporary would require at least partial pre-planning and preparation of strategies to permit their reuse and inclusion in future development or redevelopment.

3.3.2 Constraints on Effective Risk Reduction Measures in Land Use and Urban Planning

Lack of financial resources is an issue often lamented in the field of DRR and resilience. However, at some time along the disaster cycle those resources become available and therefore we must admit that constraints to more resilient settlements must be explained also on other grounds. Perhaps the most relevant reason for poor enforcement of mitigation and prevention lies in the regulatory character of planning that applies mainly to the local scale, that is in the hand of municipal governments. Local administrations are subject to the pressures constituencies put on them so they have little incentive to deny development or redevelopment in hazardous areas. In most countries larger scale schemes exist in the form of strategic or regional plans. However, the latter generally lack the teeth that are granted to local planning. In the more recent years several sectoral instruments have been enforced, for example the flood risk management plan introduced by the European Flood Directive in 2007 that is binding in several aspects. However, in the downscaling from the large catchment scale to the local jurisdiction, detailed re-evaluation of hazards and risks, structural mitigation works are proposed to loosen constraints to development. Different economic tools have been identified to support the implementation of non structural mitigation measures and primarily land use and urban planning. A well-known example of integration of insurance policies and land use planning is provided by the USA National Flood Insurance Program run by Fema and which has been the focus of extensive research and studies (May et al. 1996; Burby 2001).

A further very relevant obstacle stems from the insufficient awareness of the role played by land property rights. The impact of land ownership arrangements and regimes has been discussed in literature even though not to the extent that would be required (Platt and Dawson 1997). It is a matter that has been debated much more for developing countries rather than developed ones. However, it should be reminded that also in developed countries land ownership rights have played in the opposite direction of resilience, as developers and owners wish to make a profit developing or redeveloping hazard prone areas (Mitchell 2010). A general lack of sufficient understanding of the potential impact a natural extreme, alone or in combination, may have on the built up environment is actually shared by small owners as by large project and construction management groups that are rarely factoring prevention and resilience in their projects. Ad hoc arrangements would require to combine acquisition policies such as those carried out in Northern European Countries like Sweden with the selection of hazardous areas where public authorities wish to prohibit settlement. Studying two case studies in Australia, Harwood et al. (2014) conclude that despite rhetorical claims "creating hazard-resilient communities remains the sole responsibility of emergency management and disaster recovery authorities as these organisations deal exclusively with a community's response in the event of a natural disaster, rather than also becoming the responsibility of the land use planning and development control agencies, state/territory and local" (p. 13), and this is largely due to a legal system that substantially protects land use rights.

3.4 Conclusion

Disciplines that are searching solutions to environmental issues face significant challenges, including the need to blend different types of knowledge, manage controversial aspects of modern life, deal with contrasting interests and sometimes even with tragic choices (Calabresi and Bobbit 1978). Urban planning in particular must offer insight on the difficult balance between making cities attractive and economically vital and the requirement of keeping the impact and the footprint on ecological systems sustainable. There are ways to balance different, sometimes conflicting demands, there are tools for assessing both risks and benefits of intended investments in development, transportation, services not only for settled community but also as regards the interface between human and natural systems. Urban planning has a great advantage over other disciplines: the capacity it has matured overtime to take a systemic perspective, that permits to foresee the potential side effects of projects that look desirable but may bring unwanted impacts in the short or longer term. Yet there is the need for planners to rely on the entire set of expertise and capacities it has developed overtime, reconnecting with its tradition dating back to the XIX century when health concerns required the introduction of sanitary systems, open spaces and more balanced pattern of built surfaces permitting air and light to enter in each inhabited or occupied space (Coburn 2007). Challenges ahead relate not only to better integrate climate change mitigation and adaptation with disaster risk reduction, but also become resilient to unexpected threats so as to provide health and wellbeing to settled communities.

References

Bernstein P (1998) Against the gods: the remarkable story of risk. Wiley, New York
Bolton P, Heikkala S, Green M, May P (1986) Land-use planning for earthquake hazard mitigation: a handbook for planners. Special Publication n. 14. Boulder: Institute of Behavioral Sciences, University of Colorado, Boulder, Colorado
Bruneau M, Chang S, Eguchi R, Lee G, O'Rourke T, Reinhorn A, Shinozuka M, Tierney K, Wallace W, von Winterfeldt D (2003) A framework to quantitatively assess and enhance the seismic resilience of communities. Earthquake Spectra 19(4):733–752. https://doi.org/10.1193/1.1623497
Burby R (2001) Flood insurance and floodplain management: the U.S. Experience. J Environ Hazards 3(3):111–122. https://doi.org/10.1016/S1464-2867(02)00003-7
Calabresi G, Bobbitt P (1978) Tragic choices. W.W. Norton & Company, New York
Coburn J (2007) Reconnecting with our roots. American urban planning and public health in the Twenty first century. Urban Affairs Rev 42(5):688–713. https://journals.sagepub.com/doi/10.1177/1078087406296390
Davoudi S, Shaw K, Haider JL, et al. (2012) Resilience: a bridging concept or a dead end? "Reframing" resilience: challenges for planning theory and practice interacting traps: resilience assessment of a pasture management system. Northern Afghanistan urban resilience: what does it mean in planning practice? Resilience as a useful concept for climate change adaptation? The politics of resilience for planning: a cautionary note. Plann Theor Pract 13(2):299–333. https://doi.org/10.1080/14649357.2012.677124
Cimellaro GP, Gaudio S, Marasco S, Viapiana MF (2019) For the performance checking of road infrastructures. Territorio 89:91–96. https://doi.org/10.3280/TR2019-089012
Cimellaro GP (2016) Urban resilience for emergency response and recovery. Springer
Cutter S, Barnes L, Berry M, Burton C, Evans E, Tate E, Webb J (2008) A place-based model for understanding community resilience to natural disasters. Glob Environ Change 18:98–606
Galderisi A, Colucci A (2019) Smart, resilient and transition cities: emerging approaches and tools for a climate-sensitive urban development. Elsevier, Amsterdam
Ginzburg C (1980) Morelli, Freud and Sherlock Holmes: clues and scientific method. History Workshop 9(1):5–36. https://doi.org/10.1093/hwj/9.1.5
Greiving S, Hartz A, Saad S, Hurth F, Fleischhauer M (2016) Developments and drawbacks in critical infrastructure and regional planning. J Extreme Events 3(4):1650014. https://doi.org/10.1142/S2345737616500147
Gunderson L, Holling C (2002) Panarchy. Understanding transformation in human and natural systems. Island Press, Washington
Harwood S, Carson D, Wensing E, Jackson L (2014) Natural hazard resilient communities and land use planning: the limitations of planning governance in tropical Australia. J Geogr Nat Disas 4(2). https://doi.org/10.4172/2167-0587.1000130
Hollnagel E, Woods D, Leveson N (eds) (2006) Resilience engineering: concepts and precepts. Ashgate Publishing Limited, Aldershot, UK
Lélé S, Noorgaard R (2005) Practicing interdisciplinary. Bioscience 55(11):967–975. https://doi.org/10.1641/0006-3568(2005)055[0967:PI]2.0.CO;2
May P, Burby R, Ericksen N, Handmer J (1996) Environmental management and governance: Intergovernmental approaches to hazards and sustainability. Routledge, London
McGlade J, van den Hove S (2013) Ecosystems and managing the dynamics of change. In: Gee D, Grandjean P, Foss P, van den Hove S (eds) Late lessons from early warning. Science, Precaution and innovation. European Environment Agency, Copenhagen. https://doi.org/10.2800/70069
Melchiorri M (2016) Intergovernmental organizations and human settlements: how the world polity is shaping the debate on cities. In: Proceedings of the 52nd Isocarp Congress, Cities we have vs cities we need, 12–16 September 2016, Durban, South Africa
Margottini C, Menoni S (2018) Hazard assessment. In: Bobrovsky P, Maker B (eds) Encyclopedia of engineering geology. Springer, Cham, pp 454–478

Medd W, Marvin S (2005) From the politics of urgency to the governance of preparedness: a research agenda on urban vulnerability. J Contingencies Crisis Manage 13(2):44–49. https://doi.org/10.1111/j.1468-5973.2005.00455.x

Mejri O, Menoni S, Matias K, Aminoltaheri N (2017) Crisis information to support spatial planning in post disaster recovery. Int J Disas Risk Reduc 22:46–61. https://doi.org/10.1016/j.ijdrr.2017.02.007

Menoni S, Margottini C (eds) (2011) Inside risk: a strategy for sustainable risk mitigation. Springer-Verlag Italia, Milan

Menoni S, Bondonna C, Garcia Fernandez M, Schwarze R (2017) Recording disaster losses for improving risk modelling capacities. In: Poljansek K, Marin Ferrer M, De Groeve T, Clark I (eds) Science for disaster risk management 2017. Knowing better and losing less. European Commission, DG-JRC

Menoni S, Atun F (2017) Cities places of complexity. Educen Hanbook. Section 3.3, www.educen handbook.eu

Mitchell D (2010) Land tenure and disaster risk management. Land Tenure J 1:121–142

Nicolson C, Starfield A, Kofinas G, Kruse J (2002) Ten heuristics for interdisciplinary modelling projects. Ecosystems 5:376–384. https://doi.org/10.1007/s10021-001-0081-5

Norris F, Stevens S, Pfefferbaum B, Wyche K, Pfefferbaum R (2008) Community resilience as a metaphor, theory, set of capacities, and strategy for disaster readiness. Am J Community Psychol 41(1–2):127–150. https://doi.org/10.1007/s10464-007-9156-6

Park S, Convertino R, Linkov I (2013) Integrating risk and resilience approaches to catastrophe management in engineering systems. Risk Anal 33(3):356–367. https://doi.org/10.1111/j.1539-6924.2012.01885.x

Paton D (2008) Community Resilience: Integrating Individual, Community and Societal Perspective. In: Gow K, Paton D (eds) The phoenix of natural disasters: community resilience. Nova Science Publishers Inc, New York

Pesaro G, Minucci G, Mendoza M, Menoni S (2018) Cost-Benefit Analysis for non-structural flood risk mitigation measures: Insights and lessons learnt from a real case study. In: Proceedings of the European safety and reliability conference, Norway 17–21 June, pp 109–118

Platt R, Dawson A (1997) The taking issue and the regulation of hazardous areas. Natural Hazards Research Working Paper n. 95. Natural Hazards Research and Applications Information Center, Institute of Behavioral Science, University of Colorado, Boulder

Root T, Schneider S (1995) Ecology and climate: research strategies and implications. Science 269(5222):334–341. https://doi.org/10.1126/science.269.5222.334

Saunders W, Kilvington M (2016) Innovative land use planning for natural hazard risk reduction: A consequence-driven approach from New Zealand. Int J Disast Risk Reduc 18:244–255. https://doi.org/10.1016/j.ijdrr.2016.07.002

Schneiderbauer S, Calliari E, Eidsvig U, Hagenlocher M (2017) The most recent view of vulnerability. In: Poljansek K, Marin Ferrer M, De Groeve T, Clark I (eds) Science for disaster risk management 2017. Knowing better and losing less. European Commission, DG-JRC

Storbjörk S (2010) It takes more to get a ship to change course': barriers for organizational learning and local climate adaptation in Sweden. J Environ Planning Policy Manage 12(3):253–254. https://doi.org/10.1080/1523908X.2010.505414

Wang Q, Yamashita M (2015) Social-ecological evolutionary resilience: a proposal to enhance "sustainability transformation" about theoretical foundation. Open Access Libr J 2:e1426. https://doi.org/10.4236/oalib.1101426

Warner J, Edelenbos J, van Buuren A (2012) Space for the River Governance experiences with multifunctional river flood management in the US and Europe. IWA Publishing, London

White G, Kates R, Burton I (2001) Knowing better and losing even more: the use of knowledge in hazard management. Environ Hazards 3(3–4):81–92. https://doi.org/10.1016/S1464-286 7(01)00021-3

Chapter 4
Resilience, Cohesion Policies and the Socio-ecological Crisis

Giovanni Carrosio

The concept of resilience has been part of the public and academic debate for several years now. Since the 2008 crisis, the social sciences have borrowed from the natural sciences concepts such as fragility, resilience, persistability and adaptability (Keck and Sakdapolrak 2013). These concepts should help us to interpret and provide keys to guide action in a condition of widespread perception that many things, such as climate, economy, migration, technological innovation and so on, are out of control.

Resilience and related concepts are often used in a normative way (Wagenaar and Wilkinson 2013) because they help one imagine new socio-ecological structures that allow societies to be ready for an unstable and uncertain future. Furthermore, due to its interdisciplinary nature, the concept of resilience provides a useful tool for establishing dialogue and integration between different fields of knowledge (Beichler et al. 2014). This is very important, given the increasingly interdependent nature of the crises we are experiencing (Carrosio 2018). That does not mean, however, that the various disciplines share a common definition of resilience; significant semantic shifts have led this concept to take on different meanings. For this reason, its semantic clarity and practical relevance risk being the subject of endless debate.

According to some authors, resilience has become a boundary object (Brand and Jax 2007). Boundary objects (such as projects, ideas, maps, and texts) can easily adapt to the needs and constraints of the various parties that use them, while still maintaining a common identity across the different ways they are used. They are weakly structured in common use and become strongly structured when used by individual parties. They can either be abstract or concrete. They have different meanings according to different social settings, but their structure is common enough, in more than one world, to make them recognizable and thus a means of translation (Star and Griesemer 1989).

From this perspective, the concept of resilience evokes generic shared meanings. It is therefore capable of building a broad consensus over its use. At the same time,

G. Carrosio (✉)
Department of Political and Social Sciences, University of Trieste, Trieste, Italy
e-mail: gcarrosio@units.it

© The Author(s), under exclusive license to Springer Nature Switzerland AG 2020   49
A. Balducci et al. (eds.), *Risk and Resilience*, SpringerBriefs in Applied
Sciences and Technology, https://doi.org/10.1007/978-3-030-56067-6_4

however, precisely because it is generic, the risk exists that it will be immobilizing. When concrete actions are needed to make a territorial system more resilient, the various actors involved will be likely to disagree on what to do to achieve an apparently shared goal. Osti and Pellizzoni (2013) attributed this characteristic to the concept of sustainability. Who today would say they are in favor of the environmental and social un-sustainability of development?

However, agreeing on what sustainable development really is and what political agenda and practices must be put in place to achieve sustainability is a complicated matter. The nature of border objects has positive and negative sides. Their positive side allows researchers to build bridges between disciplines and policy makers and keep very diverse interests together under the same cognitive parameters. On the other hand, their negative side may become an obstacle to scientific and social advancement. This happens when the concept of resilience takes on overly broad and diluted connotations. In such cases, it may even hide conflicts and power relations when different people agree on the need for resilience but mean different things by it. Therefore, resilience is generally conceived as arbitrary or as an illusion and within resilience science there is confusion on how to operationalize and apply the concept (Lundgren 2020). This is why other authors have begun to study resilience as a social construct (Endress 2015).

The way to substantiate resilience depends on a series of cognitive, cultural and political frames. Frames are symbolic-interpretive constructs including beliefs, images or symbols shared by the people of any given society. Each society disposes of a limited pool of such interpretive schemes which people use to make sense of the world. According to framing theory, people tend to organize experience by relating it to previously known patterns. Perceptive elements are recognised by comparing them to a pre-existing cognitive structure (Eder 1996). We attribute different connotations to the concept of resilience and its practical variations according to the value aspects in the society-environment-technology relationship and the ways by which we perceive risks, the trust we place in technological innovation, the way we evaluate the robustness of ecosystems and, finally, how much we believe it appropriate to replace nature with artificial processes and materials.

Based on these parameters, the definition of resilience ranges from the government of uncertainty as a factor of technological innovation and new capitalist accumulation processes to seeing risk and instability factors in technological innovation and capitalist accumulation (Welsh 2012). As Pellizzoni (2017) points out, policies shifting their emphasis from the idea of mitigation to that of adaptation lead to the risk of disjunction between sustainability and resilience. Mitigation underlies theories of balance, prediction, and mitigation. Adaptation presupposes the instability of the ecosystem and is linked to ideas of uncertainty and unpredictability. Mitigation, adaptation, sustainability, and resilience are therefore not purely technical ideas. They mobilize different views on nature, society, technology, and economic growth.

To overcome this conceptual and substantial impasse, we propose to "ground" the concept of resilience. The starting point for reflection is that the social and environmental dimensions must move together, as one is a condition for the other. Environmental sustainability is a condition for welfare systems to reproduce; welfare,

Fig. 4.1 Interdependencies
in the socio-ecological crisis

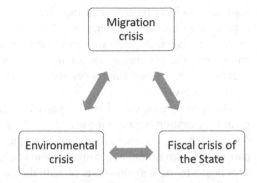

the rights of people in places, is a condition for people to be empowered in the face
of socio-ecological crises. Therefore, a resilient society can take shape from the
mix between sustainability and citizenship rights. Differently for each place, and
considering the connections between places, this balance should be built through
place-based policies. We therefore adopt a place-based approach (De Vos et al. 2019),
according to which the concept and practices of resilience should be contextualized
by places, taking into account their differences and needs. Firstly, we provide an
analytical reading of the socio-ecological crisis in its general dimension, in order to
outline the macrofactors that produce systemic instability. Secondly, we use Italy's
empty and full areas[1] to contextualize the analytical model. This is a partial and
simplified way of looking at the national territory; however, it allows us to bring out
some useful differences, from an analytical point of view, for reasoning purposes.
Starting from this contextualization, we define the risk and instability factors, and
then identify lines of intervention on which policies should act to make (empty and
full) territorial systems more resilient.

4.1   The Socio-ecological Crisis: An Analytical Model

We define the socio-ecological crisis as the intertwining of three major issues: the
fiscal crisis of the welfare state, the environmental crisis and the migration crisis
(Fig. 4.1). These three challenges are closely connected and interdependent (Gough
2010); they feed on a series of contributing factors that produce complexity and
instability on an incremental basis. In general terms, the welfare state is based on
the principle of economic growth, which should allow the state to invest more and
more in order to address demographic changes, economic crises and the promotion of inclusive development. The environmental crisis, on the other hand, shows
the contradiction between economic growth based on ever growing use of natural

---

[1]Cersosimo, Ferrara and Nisticò worked on the contrast between full and empty spaces, measuring
the density of local societies starting from four groups of indicators: physical density; demographic
robustness; economic strength and density of social relationships (2018).

resources and the finiteness of the planet. Its mitigation requires a curtailment of economic expansion in order to ward off further environmental degradation. Based on these initial elements, some researchers wonder what the future of welfare will be and refer to the environmental paradox of the welfare state (Bailey 2015). If we dig deeper, the interdependencies between welfare and the environment become much more complex.

The socio-ecological crisis took place between the late 60s and early 70s, when a great acceleration began in the use of environmental resources (Steffen et al. 2015), due to both technological innovation and the disappearance of international trade barriers. The environment became a global issue. In the same years in which sovereign debt began to grow again—to cope with the contraction of resources to be allocated to the welfare state—ecological debt[2] began to run up (Fodha and Segmuller 2014).[3] The structuring of new unfair relations between the North and the South of the world and the globalization of markets led many people to move, which accelerated the migration phenomenon. In many western countries, the migratory balance became positive in the 60s and 70s, and from that moment on, the migratory phenomenon grew and diversified, up to today's migrations, also made up of so-called environmental refugees. Interdependent environmental, fiscal and migration crises generate a greater complexity of possibilities than we are able to rationalize and manage. Their continuous feeding off each other produces strong instability and reveals the fragility of contemporary political systems.

The environmental crisis is the result of a double dynamic: acceleration and artificialisation.

Acceleration is a contradiction between the rhythms of capitalist accumulation—the speed with which nature is transformed into goods—and the time it takes for environmental resources to regenerate. O'Connor conceptualized this dynamic as the second contradiction of capitalism, capital versus environment (1991). The stronger empirical evidence of acceleration is climate change. It is determined by the growing presence of carbon dioxide in the atmosphere, which industrial societies began to emit from the mid-1800s, transforming fossil fuels into mechanical energy. With the spread of the industrial model on a planetary scale, organized on the principle of linearity, emissions have grown exponentially, reaching 400 parts per million and more (the highest value in the last 800 thousand years), considered by scientists to be a point of no return for the effects on the climate system.

Artificialisation is the principle by which the industrial system has progressed by trying to rationalize the environment, controlling and commodifying it as much as possible (van der Ploeg 2009). The end result of artificialisation is the breaking of the man-environment co-evolution nexus. Economies based on the reproduction of environmental resources become residual.

---

[2]According to Martínez-Alier (2004), ecological debt is the debt accumulated by the countries of the North towards the countries of the South in two ways: in the first place, the export of primary products at very low prices; in the second place, the free or very cheap occupation of environmental space—the atmosphere, the water, the land—through the dumping of production wastes.

[3]There are some pioneering works on the relationship between the two forms of debt, but it is a fruitful theme on which a strand of literature is developing (Boly et al. 2019).

Acceleration and rupture of the co-evolution nexus are therefore two sides of the same coin; the first leads to the formation of areas that are too full, where development is concentrated and environmental bads of anthropic origin are produced (e.g. air pollution); the second generates areas that are too empty, where territorial marginality takes shape and environmental bads (e.g. hydrogeological instability) arise from neglect. Urban systems have become global overflows, capable of responding to the needs of their citizens through the activation of global value chains. These chains organize the supply of natural resources (which are transformed into goods) on a planetary scale, disconnecting cities from neighboring areas. The marginal areas, however, have become too empty of people and too full of environment. People have migrated to cities and the environment has been abandoned: the process of co-evolution between man and the environment, which guaranteed the sustainable reproduction of ecosystems, has stopped and local environmental resources have been largely ousted from global value chains.

The fiscal crisis of the state (O'Connor 1973) is the contradiction between two functions that the state is asked to perform simultaneously: that of accumulation, to ensure that capital continues to create wealth unceasingly, and that of legitimation, to preserve social harmony (Cesareo 1978). In other words, there exists a conflict between the concentration of resources to facilitate the process of accumulation (acceleration) and the redistribution of resources to support the welfare state system. The fiscal crisis is closely connected to the environmental crisis; the end of cheap nature (Moore 2016) increases the costs of transforming nature into goods and the environmental costs of development (climate change, hydrogeological instability, air pollution, waste disposal) become a state expenditure item, which competes with the allocation of resources between capital and social rights. Environmental sustainability and sustainability of the welfare state system are therefore strictly interdependent. The global overexploitation of environmental resources weakens the possibility of guaranteeing full citizenship rights for all. On the other end of the spectrum, local underutilization of environmental resource scauses land neglect and abandonment, which becomes a cost that competes with welfare when it translates into hydrogeological instability, landslides and loss of biodiversity and ecosystem services. The process of concentration of the population around urban agglomerations has also led to the concentration of services in centres to the detriment of the most marginal areas; we therefore have areas full of services and empty of environment alongside areas empty of services and full of environment. The peripheral areas are often too full of environment because it is not managed and is in the process of degrading from a qualitative point of view.

Environmental and fiscal crises are push and pull factors for global migration. The weakening of welfare systems is a pull factor. Welfare retrenchment opens up opportunities for migrants to work in market segments not covered by the local population. On the other hand, the environmental crisis produces a massive loss of habitat due to a variety of extreme patterns, from climate change to the overexploitation of land, water and other environmental resources (Sassen 2016). The migratory phenomenon concerns empty and full areas in different ways. In general, the settlement of foreigners is linked to the dynamics of the local labour market. Therefore, it is the full areas that attract the most migrants. However, some empty areas also benefit from the presence of foreigners, thanks to specific local economies or the low cost of living.

Resilience could represent the rebalancing of sustainability in the human–environment relationship and citizenship rights (welfare services in a broad sense). According to Raworth (2017), resilience is the space between the floor of social rights and the ceiling of environmental limits (Barbera and Parisi 2019). The ceiling indicates the idea of non-surmountable borders of the biosphere; the floor, on the other hand, identifies the set of goods and access capacities that generate the basic functions of citizenship (Sen 1992). We approach this issue with a territorial posture, assuming that people have more access to services in full areas than in empty areas. We also assume that empty areas have more environment but fewer services. Let us thus consider the aggregate opportunities of people living in places. We are well aware that this is a simplification, which does not do justice to social inequalities. Independently from the territorial dimension, these contribute to generating different capacities of accessing services, which depend on the economic, relational, and cultural capital of each individual person.

4.2 The Socio-ecological Crisis in Empty Areas

Empty areas have low population density, negative demographic balances, and very high indicators of old age. They are very rich in environmental resources, which however are underutilized, with negative consequences for hydrogeological stability. Because of their social and environmental condition, they are often called fragile (Osti 2004, 2016). They are located far from the centres where the most important citizenship services are concentrated. For this reason, citizens living in empty areas have difficulty accessing services and exercising their citizenship rights. From a statistical point of view, they were identified through an indicator that measured the distance of the rural inner municipalities from the centres where the services are concentrated. The indicator was created as part of a public policy—the National Strategy for Inner Areas—which aims to combat the demographic decline of rural peripheries.[4] From a social point of view, the inner areas have low population density and have suffered from depopulation for several decades. From an environmental point of view, however, they are full of nature. However, this nature is in a state of great degradation, as a consequence of a long-populated area being abandoned.

The environmental crisis is manifested above all as abandonment of the territory, caused by a long process of "deactivation", the reduction or complete elimination of agricultural activities (van der Ploeg 2009). Abandonment does not mean recovery of a luxuriant nature, but loss of biodiversity, deterioration of ecosystems, hydrogeological instability, and increased landslide risk. Although inner areas

---

[4]For further information on the construction of the indicator and for statistical data on Italian inner areas, see Barca et al. (2018).

often enjoy a climatic advantage over full areas, they experience the acceleration dynamics, suffering the consequences of climate change in the form of extreme atmospheric phenomena amplified by environmental degradation. One such example is the cyclone Vaia, in the Belluno Alps, which destroyed thousands of hectares of forest land. There are also situations of environmental deterioration resulting from land consumption by local administrations, which, in search of economic resources to support welfare services and collect the costs of urbanization, often allow changes to the regulatory plans involving extensive cementing order. This aspect leads us to investigate the relationship between the environmental crisis and the fiscal crisis of the state. In the inner areas, the competition between the use of economic resources to cope with the growing environmental bads or to offer services to the population is increasingly tangible. Within a framework of progressive decrease in tax transfers from the centre, the tension between welfare and the environment increases the structural crisis of an unbalanced welfare system, which finds it difficult to protect the elderly while also investing in young people's needs. The migration crisis, on the other hand, is an opportunity for inner areas. Migrants entering the labour market often allow for the reactivation of society-environment co-evolution. In fact, they work mostly in agriculture, construction and pastoralism. Their presence keeps alive forms of economy linked to the active management of the territory. In addition, migrants enter the welfare system as users and service providers. As users, they contribute to maintaining the sustainability thresholds that determine the activation or suppression of a service. On the other hand, as welfare providers, they occupy parts of the labour market which are not managed by the local population. The elderly care labour market is almost totally covered by immigrants (Ambrosini 2016). Within the spiral of depopulation, fueled by demographic loss, welfare crisis and abandonment of the territory, migrants (and the neo-rural population in general) represent one of the trade-offs around which it is possible to rebuild a sustainable socio-ecological balance.

4.3 The Socio-ecological Crisis in Full Areas

Full areas have a high population density, a younger population structure and a better demographic balance than empty areas. They are heavily cemented and rich in infrastructure and industrial areas. Citizens have easy access to services, but their quality of life is undermined by air pollution, lack of green spaces and increased temperatures in the summer months, including in the form of local climate phenomena such as heat islands (Kelbaugh 2019). Soil consumption is one of the most important environmental problems in full areas, because it represents a mortgage on future land use options. Where the soil consumption percentage of empty areas is a low 4.33%, that of full area reaches as high as 12.13% (Munafò 2019). Soil consumption means failing all ecosystem services: carbon dioxide storage, water drainage, biodiversity reproduction and primary production.

In full areas, the environmental crisis takes on the connotations of acceleration and artificialisation.

Air and water pollution, soil consumption, reduction of green spaces as a consequence of soil consumption and waste production are components of acceleration dynamics. The most densely populated urban areas belong to global value chains, thanks to which they can obtain goods from the exploitation of environmental resources in different points of the planet. The imbalance between the consumption of resources and the capacity of ecosystems to absorb environmental bads results in localized (e.g. air pollution) and delocalized (e.g. landfills in empty areas) environmental degradation. The disconnection of the full areas from the surrounding environmental resources (those underutilized in the empty areas) is an element of instability and dependence of urban areas on global dynamics. At the same time, disconnection leads to the marginalization of rural areas, which do not find stable markets where to promote their natural resources.

Together with the fiscal crisis of the state, environmental degradation exacerbates inequalities and social disparities. The issue of the retrenchment of the welfare system (Korpi and Palme 2003) is in fact closely connected to the environmental crisis. The impoverishment of a part of the urban population, in particular the middle class, complicates the adoption of environmental policies designed within the market-based framework (Kete 1994). The spread of renewable energies, the retrofit of buildings and the conversion of the car fleet cannot take off due to the difficulty experienced by a large part of the population in making investments in the ecological modernization of their assets (Magnani et al. 2020). At the same time, conditions of environmental degradation affect the demand for welfare. Think of the now recognized environmental determinants of health (Gibson 2018). A growing phenomenon in full areas is energy poverty. An increasing number of citizens are struggling to pay their electricity and gas bills; others have very high default rates; still more have had their utilities cut off. Failure to use enough energy to live in healthy conditions can become an environmental determinant of health. People fall into situations of energy poverty for several reasons. The two that are most important have to do with incomes and the energy quality of the houses in which they live. The retrenchment of welfare affects the poverty conditions of people, which are becoming increasingly extreme. Even the middle class, faced with the loss of free state services, is facing difficulties in terms of energy expenditure. Housing quality affects energy requirements, which are very high for inefficient houses. There is therefore competition between household spending and government spending and between measures to reduce the energy needs of homes and welfare. Faced with declining revenues, the state is having to choose where to invest. Welfare and the environment represent items of expenditure that seem to be in competition, at least in the way they are addressed today.

Where welfare systems shrink, poverty increases and new foreign populations arrive; strong social tensions arise which result in the demand for social protection by native citizens. The competition between natives and foreigners for access to housing and services takes shape as a direct consequence of welfare retrenchment (Schuck 1998).

4.4 The Mainstream Policies on the Three Crises

To date, public policies have dealt with the environmental, migration and welfare system crises as separate areas. An interdependent reading of these phenomena did not prevail (Fig. 4.2) and this did not allow to put in place actions and policies that would have tried to reconstruct the existing relationships in order to mend the contradictions rather than exacerbate them.

The welfare crisis is tackled substantially in two ways: subtractive recalibration of services (Ferrera 2012) and retrenchment (Pierson 2001; Starke 2006). Subtractive recalibration is the adaptation of welfare to emerging social needs, however, by levelling down the performance levels on old and new social risks. Retrenchment is the reduction of spending and the use of agencies other than the state for the construction of welfare systems, such as the market and the non-profit sector. These ways of reforming welfare within the crisis, however, do not come out of the paradigm that generated it; this dynamic is entirely within the Fordist model (Koch 2013), which provided for growing economies and resources. The stable state of western economies does not allow freeing up growing resources for welfare, and the path of reduction and contextual privatization seems the only way forward. The environmental crisis is addressed within the paradigm of ecological modernization (Spaargaren 2000), which tries to solve environmental problems through economic growth-based strategies (Matlock and Lipsman 2020), such as tax policies to encourage market demand for ecological products, tax exemptions for energy saving measures, incentives for the production of energy from renewable sources and real estate policies for the enhancement of high energy class homes. None of these policies assume a social posture; for example, they do not come up with environmental solutions in such a way as to intervene on the environmental determinants of health; additionally, they are not attentive to the production and reproduction of social inequalities as a consequence of the distribution of incentives, which tend to favour the richer social classes (Metcalf 2019).

The migration issue is addressed through emergency policies and paths of subordinate integration, where a growing need for foreign workers to fill low-skilled jobs coexists with poor, residual migrant policies and a harsh political rhetoric (Ambrosini

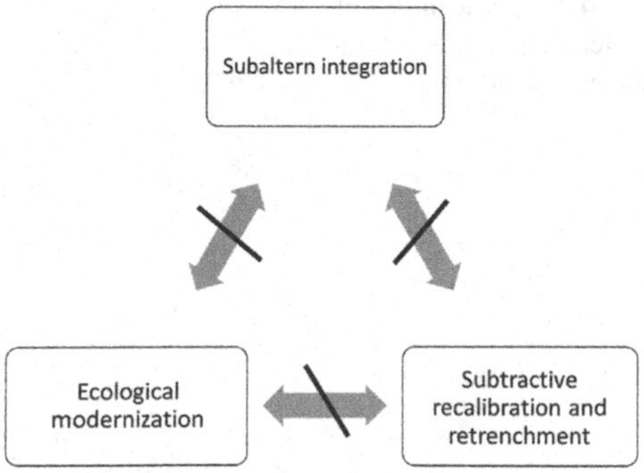

Fig. 4.2 The dominant ways in which policies intervene on the three crises

Subaltern integration

Ecological modernization

Subtractive recalibration and retrenchment

et al. 1995). The presence of immigrants is tolerated and encouraged only insofar as their role is limited to the sphere of production and social reproduction, without however recognizing full citizenship rights and entry into the sphere of political participation (Rooij 2012). This implies a weakening of the ability of immigrants to affect the social and entrepreneurial innovation of the receiving system (Bratti and Conti 2018) and to institutionalize new forms of business and welfare.

4.4.1 Interdependent Policies to Build Territorial Cohesion in the Socio-ecological Crisis

Recognizing interdependencies in the socio-ecological crisis means building political responses that look to the mending of the contradictions that feed the dynamics of the crisis. Seeking joint solutions to the environmental crisis and the fiscal crisis of the State contributes to a reconnection of empty and full areas for the sake of territorial cohesion (Fig. 4.3).

To recompose the fracture between society and the environment, it is necessary to look at forms of circular economy, which restart the co-evolution relationship between man and the environment (Kiss et al. 2019). In practical terms, this means de-artificializing some portions of the production chains by putting into circulation natural materials produced through sustainable land management. The large amount of underutilized environment in empty areas can become productive. This can be done through the production of renewable energy, natural materials for construction and clothing, foods with particular organoleptic and nutritional qualities as well as solutions for people's well-being. In order for nature to return to feeding portions of production chains in empty areas, it is necessary for institutions to work on the construction of nested markets (van der Ploeg et al. 2012) capable of protecting and remunerating natural supply chains in a fair and sustainable way. Nested markets, with their place and network specificities, can be seen as a form of hybrid governance as developed in New Institutional Economics, where specific combinations of market incentives and co-ordination modalities involving some form of hierarchical

Fig. 4.3 A new model of society born from sporadic social innovations

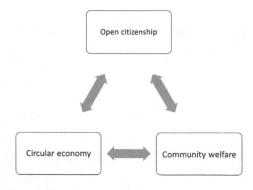

relationship characterise hybrid forms (see Ménard 1995). Nested markets should rebuild a production and consumption link between empty and full areas, leveraging on reflective urban consumption. The construction of these markets can be linked to the management and maintenance of the inner areas, as a service that local enterprises should perform in exchange for protection and greater remuneration for their products (Osti and Carrosio 2020). Within a community development framework, reactivating natural resources for production purposes can generate wealth to be used for an additional recalibration of local welfare systems, a reorganization of the welfare system that intercepts new social risks without removing rights that have already been granted. This can happen in two ways: through forms of public economy investing in local environmental resources and through new forms of fiscal solidarity introducing the remuneration of ecosystem services in the exchange between full and empty areas (Naudé et al. 2008). In this way, the organization of welfare can get out of the Fordist model of economic growth. In the debate on the transition of welfare in European countries, and in particular in Italy, a now consolidated position claims that the change in welfare regimes must go towards forms of localization and personalization (Cesareo 2017). This shift, however, implies the enhancement of all those forms of society that are between the state and the market. In many cases, the state organization is not flexible enough to modify its actions according to to the diverse needs of every place; at the same time, the market does not find the sustainability thresholds to act in places and on needs around which it is difficult to create critical mass. The connection between environment and welfare should not be created only upstream, in the way in which wealth is formed. Sewing methods can also be found downstream. For example, intervening on environmental determinants of health means making policies to combat the environmental crisis and also improve people's quality of life to decrease the need for social and health care. This can be done by building environmental policies with a social and redistributive posture. The fight against energy poverty through the retrofit of buildings, for example, allows to reduce carbon dioxide emissions and at the same time to improve the living conditions of the most vulnerable people (Grossmann 2019). People who live in better climatic conditions are less prone to getting sick during the winter. In the event that the energy retrofit interventions are made with the use of natural materials from a short chain, it is also possible to work on territorial cohesion, connecting the natural resources of empty areas with the energy saving needs of full areas. This dual posture—social and territorial—allows to build positive interdependencies between the environment and welfare, thus reversing the crisis dynamics.

Within this framework, we need to pay attention to the migration issue. If we look at those who already keep alive the economies linked to the management of natural resources and work in the field of personal care, we see how important the role of foreigners is (Baldoni et al. 2017). In fact, they fit into pieces of production chains for which it is difficult to find workforce. In addition, they already rehabilitate the empty areas, supporting the economies of scale necessary to keep population services alive. From a general demographic dynamics' standpoint, the arrival of immigrants is important because it allows to maintain essential levels of social reproduction (Lanari et al. 2020). Working on the repopulation of empty areas, with a view to future spatial deconcentration, is important, because it lightens the ecological burden of full areas (Rees and Wackernagel 1996) and supports the socio-economic dynamics of empty areas. In order for the contribution of immigrants to unfold in all its strength, however, it is necessary to overcome the subordinate integration model and imagine forms of open citizenship. Facilitating access to full citizenship rights for immigrants means taking action to counter environmental and welfare crises. Even now, immigrants regenerate the material identities of the host countries. Their contribution could be even more effective if they could enter the mechanisms of representation and democracy of the places where they live, work and produce.

4.5 Resilient Policies and Territorial Cohesion

Investigating resilience by assuming a territorial posture and setting it within the interpretative framework of the socio-ecological crisis has allowed us to clarify some analytical and interpretative elements.

First, we have identified the main elements of instability that western societies must face today and seen their complex interdependencies as contradictions within the socio-ecological crisis. Secondly, we have shown how the crisis takes shape in different areas, and consequently we have imagined resilience policies that assume a social and territorial posture.

Moving from these considerations, resilience implies the implementation of a set of policies that aim to re-balance social and territorial imbalances as a precondition for more robust socio-territorial systems.

From a theoretical point of view, therefore, we have brought the concept of resilience closer to that of sustainability and social and territorial cohesion. We believe that managing systemic instability and achieving positive resilience policies are possible only in the full exercise of citizenship rights and in situations of minor territorial inequality. Furthermore, resilience policies can become vectors for social change and therefore be desirable by citizens.

On a practical level, there are some innovations that policies could incorporate for action on social and territorial cohesion. In the first place, making sure that resilience and sustainability policies adopt criteria for improving the quality of life of the most vulnerable people, to create a connection between the environment and welfare. The example of the fight against energy poverty (as a welfare policy) through the energy retrofit of buildings (as a resilience policy) is appropriate; there may be others, for example on nutrition: working on people's diets (social determinants of health) through the enhancement of environmental resources in empty areas (reactivation and management of the area). Secondly, enhancing the eco-systemic connections between city and countryside with a view to adapting and mitigating the effects induced by climate change; social policies can become territorial cohesion policies if they work on the interdependence of city-countryside on a bio-regional scale; in terms of reducing systemic instability, this also means building paths of partial local autonomy with respect to global markets. The paths of local autonomy involve nested markets to connect reflective consumer demand with localized products; nested markets are segments of wider (more global) markets where place and network specificities provide room for specific products, extra trade, and premium prices. Nested markets imply boundaries (and often boundary organizations that actively deal with these boundaries). These boundaries can be determined by the interaction between full and empty areas, according to the logic of eco-systemic connections. Eco-systemic connections can be the vehicle for a new environmental taxation based on the remuneration of ecosystem services. Full and empty areas can be linked by a territorial solidarity pact by which full areas recognize the value of ecosystem services for empty areas. For empty areas, the possibility would open up of using this new taxation to invest in services to the population according to social investment criteria (Hemerijck 2017). Through new welfare systems that look to new inhabitants, those who are already there and those who might arrive, to manage residential mobility; a suggestion that comes to us from the recent settlement dynamics following climate change is that, in the future, residency may not be stable, but seasonal and intermittent.

References

Ambrosini M (2016) Irregular migration and invisible welfare. Springer, Berlin
Ambrosini M, Lodigiani R, Zandrini S (1995) L'integrazione subalterna. Peruviani, eritrei e filippini nel mercato del lavoro italiano. Quaderni Ismu, Milan
Bailey D (2015) The environmental paradox of the welfare state: the dynamics of sustainability. New Polit Econ 20(6):793–811. https://doi.org/10.1080/13563467.2015.1079169
Baldoni E, Coderoni S, Esposti R (2017) Immigrant workforce and labour productivity in Italian agriculture: a farm-level analysis. Bio-Based Appl Econ 6(3):259–27. https://doi.org/10.22004/ag.econ.276296
Barbera F, Parisi T (2019) Innovatori sociali. La sindrome di Prometeo nell'Italia che cambia, Il Mulino, Bologna
Barca F, Carrosio G, Lucatelli S (2018) Le aree interne da luogo di disuguaglianza a opportunità per il paese. In: Paolazzi L, Gargiulo T, Sylos Labini L (eds) Le sostenibili carte dell'Italia. Marsilio, Venezia
Beichler S, Hasibovic S, Davidse B, Deppisch S (2014) The role played by social-ecological resilience as a method of integration in interdisciplinary research. Ecol Soc 19(3):4. https://doi.org/10.5751/ES-06583-190304
Boly M, Combes J, Combes-Motel P, Menuet M, Minea A (2019) Public debt versus environmental debt: what are the relevant tradeoffs? https://hal.archives-ouvertes.fr/hal-02165453/
Brand FS, Jax K (2007) Focusing the meaning(s) of resilience: resilience as a descriptive concept and a boundary object. Ecol Soc 12(1):23. https://www.ecologyandsociety.org/vol12/iss1/art23
Bratti M, Conti C (2018) The effect of immigration on innovation in Italy. Region Stud 52(7):934–947. https://doi.org/10.1080/00343404.2017.1360483
Carrosio G (2018) I margini al centro. L'Italia delle aree interne tra fragilità e innovazione. Donzelli, Roma
Cersosimo D, Ferrara A R, Nisticò R (2018) L'Italia dei pieni e dei vuoti. In: De Rossi A (eds) Riabitare l'Italia. Le aree interne tra abbandoni e riconquiste. Donzelli, Roma
Cesareo V (1978) Cittadinanza e Stato assistenziale. Studi di Sociologia 16(3/4):279–301

Cesareo V (2017) Per un welfare responsabile. In: Cesareo V (ed) Welfare responsabile. Vita e Pensiero, Milano

de Rooij E (2012) Patterns of immigrant political participation: explaining differences in types of political participation between immigrants and the majority population in western Europe. Eur Sociol Rev 28(4):455–481. https://doi.org/10.1093/esr/jcr010

De Vos A, Biggs R, Preiser R (2019) Methods for understanding social-ecological systems: a review of place-based studies. Ecol Soc 24(4):16. https://doi.org/10.5751/ES-11236-240416

Eder K (1996) The institutionalisation of environmentalism: ecological discourse and the second transformation of the public sphere. In: Lash S, Szerszynski B, Wynne B (eds) Risk, environment and modernity. Sage, London

Endress M (2015) The social constructedness of resilience. Soc Sci 4:533–545. https://doi.org/10.3390/socsci4030533

Ferrera M (2012) Verso un welfare più europeo? Conclusione. In: Ferrera M, Fargion V, Jessoula M (eds) Alle radici del welfare all'italiana. Marsilio, Venezia

Fodha M, Seegmuller T (2014) Environmental quality, public debt and economic development. Environ Resour Econ 57:487–504. https://doi.org/10.1007/s10640-013-9639-x

Gibson JM (2018) Environmental determinants of health. In: Daaleman T, Helton M (eds) Chronic Illness Care. Springer, Cham

Gough I (2010) Economic crisis, climate change and the future of welfare states. Twenty-First Century Society 5(1):51–64. https://doi.org/10.1080/17450140903484049

Grossmann K (2019) Energy efficiency for whom? A conceptual view on retrofitting, residential segregation and the housing market. Sociologia Urbana E Rurale 119:78–95. https://doi.org/10.3280/SUR2019-119006

Hemerijck A (2017) The uses of social investment. Oxford University Press, Oxford

Keck M, Sakdapolrak P (2013) What Is social resilience? Lessons learned and ways forward. Erdkunde 67:5–18. https://doi.org/10.3112/erdkunde.2013.01.02

Kelbaugh D (2019) The urban fix. resilient cities in the war against climate change, heat islands and overpopulation. Routledge, New York

Kete N (1994) Environmental policy instruments for market and mixed-market economies. Util Policy 4(1):5–18. https://doi.org/10.1016/0957-1787(94)90030-2

Kiss K, Ruszkai C, Takács-György K (2019) Examination of short supply chains based on circular economy and sustainability aspects. Resources 8:161. https://doi.org/10.3390/resources8040161

Koch M (2013) Welfare after growth: theoretical discussion and policy implications. Int J Social Quality 3(1):4–20. https://doi.org/10.3167/IJSQ.2013.03010

Korpi W, Palme J (2003) New politics and class politics in the context of austerity and globalization: welfare state regress in 18 countries, 1975–95. Am Polit Sci Rev 97(3):425–446. https://doi.org/10.1017/S0003055403000789

Lanari D, Pieroni L, Salmasi L (2020) Regularization of immigrants and fertility in Italy. MPRA Paper No. 98241, posted 23 Jan 2020

Lundgren J (2020) The grand concepts of environmental studies boundary objects between disciplines and policymakers. J Environ Stud Sci. https://doi.org/10.1007/s13412-020-00585-x

Magnani N, Carrosio G, Osti G (2020) Energy retrofitting of urban buildings: a socio-spatial analysis of three mid-sized Italian cities. Energy Policy 139:111341. https://doi.org/10.1016/j.enpol.2020.111341

Martínez-Alier J (2004) El ecologismo de los pobres. Confl ictos ambientales y lenguajes de valoración. Icaria–FLACSO, Barcelona

Matlock AS, Lipsma JE (2020) Mitigating environmental harm in urban planning: an ecological perspective. J Environ Planning Manage 63(3):568–584. https://doi.org/10.1080/09640568.2019.1599327

Ménard C (1995) Markets as institutions versus organizations as markets? Disentangling some fundamental concepts. J Econ Behav Organ 28:161–182. https://doi.org/10.1016/0167-2681(95)00030-5

Metcalf G (2019) The distributional impacts of U.S. energy policy. Energy Policy 129:926–929. https://doi.org/10.1016/j.enpol.2019.01.076

Moore Jason W (2016) The rise of cheap nature. Sociology Faculty Scholarship 2. https://orb.bin ghamton.edu/sociology_fac/2

Munafò M (2019) Consumo di suolo, dinamiche territoriali e servizi ecosistemici. Edizione 2019. Report SNPA 08/19

Naudé A, Le Maitre D, De Jong T, Forsyth G, Mans G, Hugo W (2008) Modeling of spatially complex human-ecosystem, rural-urban snd rich-poor interactions. Paper submitted to the international conference studying, modelling and sense making of planet earth. Department of Geography, University of the Aegean, 1–6 June 2008

O'Connor J (1973) The fiscal crisis of the State. Martin Press, New York, St

O'Connor J (1991) On the two contradictions of capitalism. Capitalism Nature Socialism 2(3):107–109

Osti G (2004) Un'economia leggera per aree fragili. Criteri per la sostenibilità ambientale nel Nord Italia. Sviluppo Locale 11(27):9–31

Osti G (2016) The unbalanced welfare in Italian fragile rural areas. In: Grabski-Kieron U, Mose I, Reichert-Schick A, Steinfurher (eds) A European rural peripheries revalued. Governance, actors, impacts. Lit Verlag, Münster

Osti G, Carrosio G (2020) Nested markets in marginal areas: weak prosumers and strong food chains. J Rural Stud 76:305–313. https://doi.org/10.1016/j.jrurstud.2020.04.004

Osti G, Pellizzoni L (2013) Sociologia dell'ambiente. Il Mulino, Bologna

Pellizzoni L (2017) I rischi della resilienza. In: Mela A, Mugnano S, Olori D (eds) Territori vulnerabili. Verso una sociologia dei disastri italiana. FrancoAngeli, Milano

Pierson P (2001) The new politic of the welfare state. Oxford University Press, Oxford

Ploeg van der JD (2008) The new peasantries. Struggles for Autonomy and Sustainability in an Era of Empire and Globalization. Routledge, London

Raworth K (2017) Doughnut economics: seven ways to think like a 21st-century economist. Random House, New York

Rees W, Wackernagel M (1996) Urban ecological footprints: why cities cannot be sustainable—and why they are a key to sustainability. Environ Impact Assess Rev 16:223–248. https://doi.org/10.1016/S0195-9255(96)00022-4

Sassen S (2016) A massive loss of habitat. New drivers for migration. Sociol Dev 2(2):204–233. https://doi.org/10.1525/sod.2016.2.2.204

Sen AK (1992) Inequalities reexamined. Oxford University Press, Oxford

Schuck P (1998) Citizens, strangers and in-betweens. Routledge, New York

Spaargaren G (2000) Ecological modernization theory and domestic consumption. J Environ Planning Policy Manage 2(4):323–335. https://doi.org/10.1080/714038564

Star SL, Griesemer JR (1989) Institutional ecology, 'translations' and boundary objects: amateurs and professionals in Berkeley's Museum of vertebrate zoology, 1907–39. Soc Stud Sci 19(3):387–420. https://doi.org/10.1177/030631289019003001

Starke P (2006) The politics of welfare state retrenchment: a literature review. Soc Policy Admin 40:104–120. https://doi.org/10.1111/j.1467-9515.2006.00479.x

Steffen W, Broadgate W, Deutsch L, Gaffney O, Ludwig C (2015) The trajectory of the anthropocene: the great acceleration. Anthropocene Rev 2(1):81–98. https://doi.org/10.1177/2053019614564785

van der Ploeg JD, Jingzhong Y, Schneider S (2012) Rural Development through the construction of new, nested, markets: comparative perspectives from China, Brazil and the European Union. J Peasant Stud 39(1):133–117. https://doi.org/10.1080/03066150.2011.652619

Wagenaar H, Wilkinson C (2013) Enacting resilience: a performative account of governing for urban resilience. Urban Stud 52(7):1265–1284. https://doi.org/10.1177/0042098013505655

Welsh M (2012) Resilience and responsibility: governing uncertainty in a complex world. Geogr J 180(1):15–26. https://doi.org/10.1111/geoj.12012

Chapter 5
Some Notes on Socio-economic Territorial Imbalances in Contemporary Italy

Gianfranco Viesti

5.1 Introduction

To understand the size, the characteristics and the evolution of socio-economic territorial imbalances within Italy, one has necessarily to take into account three main trends of the country as a whole in the twenty first century. First of all, Italy underwent a profound demographic transition: fertility rates fell well below the replacement ratio, so that natural changes became negative; in the same time Italy became a country of immigration after a long history as a country of emigration; and internal migration remained quite high.

Second, overall economic development was negative: in the twenty first century Italy has faced the longest and deepest period of economic recession of its whole history. Economic growth was limited before the international and then eurozone economic crisis of 2008–09 and 2011–12; during the crisis production and income per capita fell; the recovery in the last years of the second decade of the century was modest. Overall economic performance was weaker than in other European countries. Third, public policies were reduced and changed, due to both the need of austerity in public finance and the diffusion of neo-liberal prescriptions for economic policy. These trends have had asymmetrical effects on Italian regions and cities: territorial imbalances, already large, were affected.

What follows is therefore organized to give a few hints on the impact of the new situation of the country on its territorial imbalances. So, Sect. 5.2 describes demographic changes and their consequences. Sections 5.3, 5.4 and 5.5 are devoted to economic developments: Sect. 5.3 to regional disparities; Sect. 5.4 to employment and Sect. 5.5 to the a more detailed territorial analysis (local labor systems).

G. Viesti (✉)
Department of Political Sciences, University of Bari, Bari, Italy
e-mail: gianfranco.viesti@uniba.it

Section 5.6 briefly deals with the changes in some of the most important public policies and concludes, also taking into account the political approaches to territorial
disparities.

5.2 Demographic Changes

The demography of Italy has changed and is still changing. Italy was traditionally
a quite prolific country, while nowadays is one of the world's country with lower
fertility rates; Italy was traditionally a country of emigration, while nowadays is
a country of immigration; on the contrary, Italy remains a country with internal
migrations from the South to the North and from the countryside to the cities. These
phenomena have relevant consequences for territorial disparities.

In 1964, in the year of largest increase of population due to the baby boom, a
million new Italians were born; the more recent figure is around 450,000 (Golini
2019). The demographic transition has had a different timing in the two large macro-
areas inside Italy: in the Centre-North it started (in particular with the decrease of
fertility rates) well before, so that until the end of the twentieth century fertility rates
and the rate of growth of population were higher in the South. In the twenty first
century those rates have converged: fertility rate is now well below the replacement
rate (2.1 children per woman) everywhere in the country, and the natural rate of
increase of the population has become negative, for the first time in centuries.

In the same time, starting from the end of the Nineties, there were large inflows of
immigrant population. However, being migrations mostly due to economic reasons
(Coniglio 2019), migrants went in search of employment opportunities; that were
much larger in the Centre-North. The presence of foreign-born population is therefore much higher in the Centre-North than in the South. Immigration from abroad
improved the demographic decline in the Centre-North, being immigrants younger,
on average, than resident population. As an important consequence, almost half of
the new children born in the Centre-North in the last ten years were of foreign
nationality. Starting from 2010 immigration from abroad to the whole of Italy is
decreasing; and a new flow of Italian people migrating abroad is now registered, due
to the economic stagnation of the country. The new Italian emigration is still small
(even if the numbers might be under-reported: see Pugliese 2018) but increasing.
Migrants come both from more developed regions of the North and from less developed regions of the South. Notwithstanding decreasing inflows and the new outflow
of Italian people going abroad, the net contribution of international migration to
population is still positive. Internal migrations from the South to the North, and from
the countryside towards the urban areas remain a feature of the Italian society; they
are much smaller than in the decades of larger movement of population, such as the
1960's, but still significant.

As a result of the natural change of population, and of the international and internal
migrations, in the twenty first century the demographic balance was very different in
the two macro-areas: in the Centre-North there was a natural decrease of 0.7 million

(2002–16), but a migratory positive balance of 4.1 million (3.3 from abroad, 0.7 from the South) with an overall balance of + 3.4 million; on the contrary, the South experienced a minimal increase, due to the internal out-migration and the relatively limited immigration from abroad (0.8 million). However, during the period the situation worsened, so that in more recent years the population is actually decreasing. In smaller municipalities the demographic decrease was more intense (Svimez 2019): in the Mezzogiorno, the population of municipalities with less than 5,000 inhabitants decreased by 250,000 (2003–2017); in the Centre-North, the population of smaller municipalities in mountain areas decreased as well.

The reduction and aging of population has deep socio-economic consequences: for the magnitude of the internal demand; for the ratio of labor force over total population; for the need of services for the elderly, for the size and the opportunities of work in the more important public services: for example, school population in Italy is projected to decrease by 16% (2019–2030) (Fondazione Agnelli 2020), but the decrease will be much larger in the South, where half of the total reduction of students is likely to occur.

Table 5.1 tries to summarize the demographic changes (2006–2018), presenting the population trends for macro-regions and local areas. The Italian National Institute of Statistics (ISTAT) provides data on population and employment at a quite detailed territorial dimension. That is, the 610 local labor system (LLS) in which Italy can be divided, according to a methodology that takes into account both registered residence and home-work daily movements. For the goals of this paper, the 610 have been divided into three main groups: large urban areas (LLS having a population of more than half a million), intermediate areas (LLS with a population between 100 and 500 thousands), and smaller areas (less than 100.000).

The data of Table 5.1 confirm that the demographic change has been more positive in the Centre-North than in the South, but also shows that population increased much more in urban (5.9%), compared to intermediate (3.7%) and smaller areas (0.6%). The increase of the population, both in absolute and relative terms, was particularly remarkable in Rome and Milan. However, all urban areas in the Centre-North registered a demographic growth, larger than the national average. With the notable exceptions of Turin and Genoa: the latter was the only Italian urban LLS that

Table 5.1 Demographics, 2006–2018 (percentage change of total population)

	Large (1) LLS	Intermediate (2) LLS	Small (3) LLS	Total
North	5.7	5.2	2.3	4.6
Centre	10.9	5.8	1.6	6.7
South	2.3	0.3	−1.3	0.2
Italy	5.9	3.7	0.6	3.5
(1) More than 500,000 inhabitants				
(2) Between 100,000 and 500,000 inhabitants				
(3) Less than 100,000 inhabitants				

Source Elaboration on ISTAT data

actually had a demographic decline. As we will see later, there are interesting social and economic differences within the Centre-North. Population increased in all urban systems in the South, though less than the national average; smaller Southern LLS actually registered a demographic decline.

5.3 Regional Disparities

The main feature of the recent Italian development was the economic negative performance, compared to the EU, in all its regions. The Italian economy is having a dismal performance in the twenty first century: its rates of growth are much lower than the European average. Growth was weaker before the economic crisis; the recession due to the financial and sovereign debt crises was harder; the recovery in more recent years slower than in other EU Member States. All Italian regions lost ground compared with the EU average. The GDP per capita at purchasing power parity of the old industrial region of Piedmont in the North was in 2017 only slightly above the EU-28 average (it was 18% above in 2006); even Lombardy (the largest and economically stronger area of the country) went in the same period from an index of 138 (EU-28 = 100) to 128.

It is not in the goals of these notes to explain the main reasons behind this national performance. However, some of the main causes of the relative "decline" of the Italian economy, as suggested in the relevant literature, can be mentioned, as they will be used in the territorial analysis. Italy suffered both the commercial competition of emerging economies, especially China, in more traditional industries and the locational competition of Central and Eastern European countries in attracting foreign direct investments, especially from Western Europe in the more labor-intensive tasks of global value chains (Viesti 2019c). The substitution of declining firms and industries with more innovative and productive ones was more difficult due to the traditional weakness of Italian industry in RD-intensive and innovation-intensive manufacturing and services. The country also showed a relative underdevelopment of new service activities, especially those based on digital technologies, with their high skill job demand. Public policies did not help, due to underinvestment, especially in the 2010s in both material and immaterial infrastructures. Further research is needed to demonstrate how regional differences in these regards help explaining territorial performances; some tips will be given in next pages.

Italy has traditionally quite large regional disparities (Iuzzolino et al. 2013); GDP per capita is much smaller in the eight regions of the South than in the North; the regions of the Centre are in between. In the decade 2007–2017, when Italian GDP decreased by 5.2%, those disparities increased: the reduction of economic activity was smaller in the North (−2.3%) than in the Centre (−6.9%) and in the South (−10.4%) (Banca d'Italia 2019). In terms of the contribution of main industries to the overall performance, construction plummeted everywhere (the national value added decreased more than 31% from 2007 to 2017); services did not grow particularly, but Northern regions showed a better than the average performance, while Southern regions performed worse; this appears to be a key difference. Another important territorial difference was in the manufacturing industry; it lost 10% of its value added as a national average, but the decline was smaller in the North than in the South; the latter lost around one quarter of its production. Technological developments, the increase of import from emerging economies and the reorganization of the European industry around Germany and Central European economies, as mentioned (Viesti 2019a) hit more the relatively weaker Southern industry, characterized by a larger presence of smaller and less innovative firms.

While Italian GDP declined by 5.2% from 2007 to 2017, a small group of regions performed relatively better (Table 5.2). They are the three large regions of Lombardy, Emilia-Romagna and Veneto, together with the autonomous provinces of Bolzano and Trento (whose economies follows German economic cycles much more than the Italian one). Toscana follows, relatively close. All other regions performed even worse than the Italian average, with the exception of the two small and highly industrialized Southern regions of Abruzzo and Basilicata). They include areas in both the North West (Piedmont, Liguria) and the North East (Friuli Venezia Giulia): in those regions the decline of manufacturing was not matched by a rise in new professional services. Two of the Central regions (Marche and Umbria) performed particularly badly. All large Southern regions recorded negative trends, around 10% of their production.

The part of the Italian economy showing a better resilience resembles to a Cross; it stretches North–South from the Austrian border at the Brenner pass through Verona and Bologna to Florence; West–East from Milan through Bologna up to Rimini, and through Verona up to Treviso. An area characterized by relatively strong and well integrated urban economies.

In terms of GDP per capita there are some differences with the previous picture, due to the different dynamics of population. The relative decline of some of the Northern and Central regions appears even stronger, because they saw an increase of population together with a decline of GDP. Southern regions still perform worse than the national average, even if the difference becomes smaller.

Table 5.2 Regional GDP growth 2007–17

North-West 2.1	North-East −2.5	Centre −6.9	South −10.4
Piedmont −8.2	Bolzano 9,0	Toscana −4.5	Abruzzo −5.1
Aosta Valley −11.8	Trento 0.6	Umbria −15.6	Molise −20.1
Lombardy 1.7	Veneto −3.4	Marche −11.6	Campania −11.5
Liguria −11.2	Friuli V.G. −7.8	Lazio −6.1	Puglia −7.6
	Emilia-Romagna −2.1		Basilicata −0.1
			Calabria −12.8
			Sicilia −13.2
			Sardegna −9.7

Source Banca d'Italia (2019)

5.4   Employment Trends

Regional disparities depend upon differences in productivity and employment rates. The much lower GDP per capita of the Southern regions is due to both causes: their GDP per capita is only 55% of that of the Centre-North, due to a double gap in employment rates (around 30% smaller) and productivity (around 20% smaller). The role of the employment gap to explain regional disparities is in Italy (ad well as in Spain) larger than in other advanced countries (OECD 2018).

However, productivity and employment are in part two sides of the same coin: economic growth comes from the rise of employment in the more productive segments of the economy in both manufacturing and service activities; it is a matter of both industrial composition ("between sectors") and of the presence of better performing firms within each industry ("within sectors"). We have interesting and detailed data on employment. Again, we need to analyze regional and local situations keeping in mind national trends; remembering, as well, that in last years the number of part-time jobs (and in particular of involuntary part-time jobs) and the share of employed on temporary basis increased in Italy. To understand national and regional trends the comparative analysis of regional and national labor markets in the EU performed by the JRC-Eurofound (2019) seems particularly useful. The JRC analysis presents both overall and sectoral dynamics of jobs; it adds, for the first time, an analysis of "quality of jobs". JRC-Eurofound presents the change of employment divided between wage terciles: crossing professions and industrial JRC is able to create three groups of jobs in terms of their relative wage: low-paid, mid-paid and high-paid. In reading the data one has to remember that the division of total employment in terciles is based on the average of nine European countries in different years, so that national and regional changes must be read as compared to the general trends, that see an improvement—in terms of quality of jobs as measured by wage—of employment.

JRC-Eurofound (2019) shows that Italy experienced employment growth mostly in low-paid jobs, in the service sector; the country lost jobs in the group of professions and occupations with intermediate wage levels (both in manufacturing and construction), with a modest increase in the better paid jobs (again in services). This national pattern contrasts with more positive trends recorded in the other larger European countries, even if different between them: with France showing a polarization on employment (growth both in low-paid and high-paid jobs) and Germany an increase in the jobs with intermediate wages.

Together with the general economic trends, an explanation of this performance comes from the changes in the sectoral composition of employment between 2002 and 2017 (JRC-Eurofound 2019, Table 1). In Italy the weight of manufacturing decreased, as happened in all other Western European countries. However, it remains substantial, with an employment share of 18%, that is slightly lower than in Germany and much higher than in France or the United Kingdom. What appears to be more important is that: (a) Italy shows lower shares of total employment in public administration, education and health, with the weight of public administration scoring a marked decrease; those are activities characterized by mid-paid and high-paid jobs; and that: (b) sectors with increasing shares of overall employment were polarized: on the one side (as in all European countries) there was a growth of relatively high-paid jobs in professional services; on the other side relatively low paid jobs in "accommodation and food", due to the development of tourism, and in "household as employees", mostly immigrant workers. Having "household as employees" is typical of Southern European countries such as Italy and Spain, where the Mediterranean version of the welfare states does not provide families with services for the elder and the youngest, as in the rest of the EU.

These changes in industrial shares of total employment matter for territorial disparities. There was a composition effect. Public administration, education and health are territorially diffused, with relatively well paid jobs in all regions. On the contrary the geographical polarization of jobs in professional services is even higher than in manufacturing: they are concentrated in the strongest urban areas.

Consequently, regional trends in employment, as measured by JRC-Eurofound (2019) show contrasting patterns (Fig. 5.1); the best performing Italian regions (Lombardy, Emilia and Veneto) have an overall increase of employment (2002–17): their new jobs are both at intermediate wage levels (probably also due to their resilience in manufacturing) with a modest increase in the relatively high-paid jobs (professional services). Other Northern regions, such as Piemonte and Liguria actually see a decrease of the number of high-paid jobs, even if in the former mid-paid jobs increased. Lazio shows an original trend, with an increase of jobs at all levels, but particularly in low-paid jobs (likely due to tourism in Rome). All Southern regions show disappointing trends, with an absolute decrease in the better paid tercile of

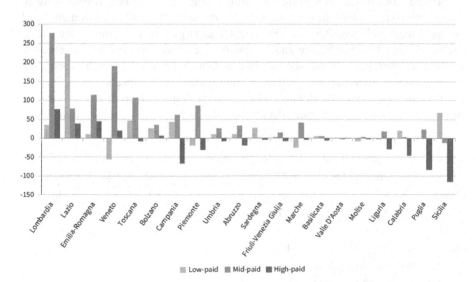

Fig. 5.1 Absolute change in jobs, by region and tercile (compared to the average in 9 Member States), 2002–2017. *Source* JRC-Eurofound (2019)

jobs (mostly public jobs): they have "downgraded" their labor market compared to Europe (JRC-Eurofound 2019, Fig. 26).

5.5 Cities and Local Labor Systems

Together with disparities between regions, the Italian economy shows interesting, and growing, differences within regions. To track those features, it is useful to give a look at the data on the employment performance (2006–18) of the 610 Italian local labor systems (LLS). We will compare their performance with the national increase in the decade: 2%, that is 457,000 units.

Main results of the territorial analysis can be summarized in the following main points. First, larger urban areas performed much better than the rest of the country (Table 5.3).

In the 16 LLS with a population greater than half a million, that host one third of the total Italian population, there was an increase of 303.000 jobs (+4.0%), the double of national average. The 112 LLS with a population between 100,000 and 500,000—that represent another third of total population—marked an increase of 133,000 jobs (+1.7%). The 482 smaller LLS a very modest increase (+0.3% or 21,000 jobs). Only 241 LLS out of the total 660, even if representing one half of total population, performed better than national average, against 369 with a worse performance. Among the latter, 294 experienced a decline of employment; in 38 of them, all but one in Southern Italy, the decline was larger than 10%.

Second, not all the cities performed well (Table 5.4). The best results were in Rome, as well in some of the largest urban systems of the "Cross" that was mentioned before: Milan, Bologna, Bergamo, Firenze. Other large urban systems of the Cross, such as Como, Padova, Venice and Bergamo, had a more modest but nonetheless better than average performance. Turin and Genoa performed much worse. All Southern urban areas show a decline of employment, that appears quite strong in the two Sicilian areas of Palermo and Catania. It is not only a matter of the size of population; employment rates decline as well in Southern urban systems. As a

Table 5.3 Employment change, 2006–2018 (percentage change of total population in LLS)

	Large (1) LLS	Intermediate (2) LLS	Small (3) LLS	Total
North	4.3	3.8	2.9	3.7
Centre	13.0	4.2	3.9	7.4
South	−6.3	−4.5	−4.3	−4.9
Italy	4.0	1.7	0.3	2.0

(1) More than 500,000 inhabitants

(2) Between 100,000 and 500,000 inhabitants

(3) Less than 100,000 inhabitants

Source Elaboration on ISTAT data

Table 5.4 Employment growth in the largest urban labor systems, 2006–2018

	Region	Area	Absolute change	Percentage change	Employment rate	
					2006	2018
ROMA	Lazio	Centre	203.4	14.6	48.8	49.2
MILANO	Lombardy	North-West	116.1	7.2	52.2	51.8
BOLOGNA	Emilia	North-East	26.2	7.0	52.4	53.3
FIRENZE	Toscana	Centre	16.3	5.6	49.2	49.1
COMO	Lombardy	North-West	10.6	4.5	53.5	52.0
PADOVA	Veneto	North-East	11.8	4.1	52.4	50.6
VENEZIA	Veneto	North-East	9.8	4.0	47.6	47.9
BERGAMO	Lombardy	North-West	13.8	3.9	54.9	51.7
BUSTO ARSIZIO	Lombardy	North-West	3.7	1.3	53.9	51.4
TORINO	Piedmont	North-West	−0.8	−0.1	48.1	47.0
GENOVA	Liguria	North-East	−5.3	−1.9	44.6	45.7
CAGLIARI	Sardinia	South	−4.7	−2.5	43.6	40.8
BARI	Puglia	South	−7.3	−3.1	38.8	36.0
NAPOLI	Campania	South	−43.9	−6.1	35.2	31.4
CATANIA	Sicily	South	−19.1	−9.1	37.7	31.9
PALERMO	Sicily	South	−27.2	−10.4	36.1	30.7

Source Elaborations on ISTAT data

result (Table 5.3, again), contrary to what happened in the country as a whole, and in the Centre-North, Southern larger LLS's did not perform better than the rest of the macro-region. It appears to be a serious problems of Southern cities: they lost jobs in public administration, health and education; in some cases they had a contraction of their manufacturing base; in the same time they were not able to create new jobs in more advanced private services.

Third, industrialization per se does not seem to explain the difference. In the 141 LLS's labelled by ISTAT as "industrial districts", employment growth between 2006 and 2018 was 2%, the same rate of the country as a whole. Interestingly, and coherently with the national picture sketched before, in the 50 LLS's labelled by ISTAT as "touristic", the increase was larger, up to 3.6%. On the contrary, several more developed LLS, especially in the North-West and in the Centre, and in particular on the Adriatic coast, in the Marche region had negative performance. All those more detailed trends deserve further research.

The relative performance of the cities (using provincial data) is also visible from the data describing net migration flows of Italian young people (25–39 years) with tertiary degrees, and the percent of young people having tertiary degrees (Table 5.5).

In reading the table, one has to take into account that Italy is experiencing a growing outflow of tertiary graduates, coming from all the regions. The migration balance of the most important provinces inside the country confirms an increasing

Table 5.5   Young people with tertiary degrees in main Italian provinces

	% of young people (24–39 years) with tertiary degrees	Balance of mobility of young people with tertiary degrees	
	2016	2004	2016
TORINO	26	7	0
GENOVA	28	7	−1
MILANO	36	17	35
BOLOGNA	37	29	33
PADOVA	28	5	−6
FIRENZE	37	10	5
ROMA	31	15	0
NAPOLI	19	−20	−22
BARI	23	−15	−18
CATANIA	17	−6	−23
PALERMO	17	−15	−26
CAGLIARI	19	−4	−10

Source Elaborations on ISTAT data

attractive role of Milan and Bologna; other provinces (not in the table) with positive and growing balances are those located between the two cities, mostly in Emilia-Romagna. On the contrary, the other main cities of the Centre-North (with the exception of Florence) seem to have lost their attractive power, with their net balance of incoming vs. outgoing tertiary graduates decreasing. Southern cities confirm their role as "exporters" of qualified young people, with an increase of outflows especially for the Sicilian cities of Catania and Palermo. Rural Italy, in the North and to a greater extent in the South (not in the table), sees a substantial outflow of young and qualified human capital: provinces such as Enna in Sicily or Foggia in Apulia are experiencing a yearly outflow around 4% of their stock of young tertiary graduates. The Mezzogiorno lost 117,000 young graduates in the last ten years (2008–2017) (ISTAT 2019).

As a result of both different rates of university enrollment and graduation (Viesti 2018) and migration of university students and tertiary graduates, the percentage of young (25–39) people holding a tertiary degree goes from 36–37% in Milan, Bologna and Florence, to 31% in Rome, 26–28% in Padua, Turin and Genoa, down to 23% in Bari, 19% in Naples and Cagliari and 17% in Catania and Palermo. It clearly appears a very different potential of Italian cities in the supply of the human capital needed for the development of the new digital economy.

Two more dimensions of territorial disparities should be added. The first with reference to the trends affecting Italian "internal areas", that is areas further away from main cities, with a decreasing availability of health, education and mobility services (De Rossi 2018). Several internal areas are losing population, and this may put at risk their sustainability in the next future. The second with reference to disparities within larger urban areas, with a growing divide between city centers and peripheries. Both phenomena deserve to be studied.

A general conclusion of this analysis is that, while the whole country is experiencing a quite negative period, some cities and regions, the before mentioned "cross" seem to perform relatively better; internal disparities increased; and might increase even more in the future, given the present trends.

5.6   Public Policies

In the twenty first century, especially in the second decade, Italian public policies have not been successful in reducing those disparities; to some extent they seem to have contributed to their enlargement. This was due to overall fiscal policies as well as to the political choice to concentrate public intervention in the better performing areas. In this paragraph, some examples will be given.

Italian budget policy went into a period of prolonged austerity in the 2010's. This was due to the need of reducing the imbalances of Italian public finances (in terms of public deficit and debt), following EU budget rules, after the international and then the euro crises. Italian austerity was particularly incisive in the 2011–2015 period.

Italian austerity appears to have been asymmetric among people and regions. First, Italy experienced growing social disparities, even if to a minor extent than in other advanced countries; inequality increased both in terms of income and wealth: in 2006–2016 the decrease of the income of the poorest 10% of the country was much larger than average (Banca d'Italia 2019). But the composition of population in different areas is not the same: poorer areas have a larger than average presence of poorer people; so that increasing social disparities may have implied an increase in territorial disparities, as Manduca (2019) demonstrates for the United States.

Fiscal policies hit relatively more weaker parts of the country. An short illustration of this trend comes from the dynamics of public expenditures: in 2001–2018 they increased by 11.6% in the Centre-North and decreased by 2.4% in the South (Svimez 2019). This was due to the change in the composition of public expenditures: in particular to the decrease of public investment; and to discretionary choices regarding specific policies.

Capital expenditures were reduced much more than current expenditures. In particular, public investment reached a very low level after 2009–10; capital incentives to private investment reached as well a historical minimum (Svimez 2019). Both figures, in terms of percentage of GDP were, and still are, well below the EU average. This is particularly important for regional development policies; capital expenditures are important for all areas, but they are particularly relevant for less developed areas, whose infrastructural equipment is below the national level, especially in Italy. Total capital expenditures in the Mezzogiorno were until the end of the first decade of the century around 20 billion euros; they went to 10 billion euros (in real terms) in 2017–18. This implies that European (structural funds) and national regional policy expenditures in Italy are only partially compensating the decline of overall public investment (Viesti 2016).

As an example of the choices regarding the allocation of public investment, one can take high-speed rail. Italy was able to create in the twenty first century a very successful network of high-speed rail; the new tracks allowed high quality new transport services. However, they cover only the most developed parts of the country: basically the "Cross", up to Salerno in the South. The network left aside the weaker parts of the North and the Centre, and almost all the South (Viesti 2019b). Moreover, in those regions the railways services actually decreased due to the budget cuts in their financing. The political decision, basically, was that public policies should follow the regional development, creating new advanced transport services only where demand already exists, due to higher income level; renouncing to the decisive role of public investment and transport services to ease economic development. Investing in advanced services in more developed areas is an appropriate choice to support and increase their competitiveness; but investing only there implies a large increase in territorial disparities. Total railway public investment per capita, 2000–17, was 58 euros per year in the South, as compared to 116 in the North-West.

Large current expenditure policies were affected as well. An important example of discretionary choices are the ones concerning universities (Viesti 2018), whose role is crucial for territorial development. Starting in 2008 total expenditures for tertiary education, already at a low level in Italy in the international comparison, were reduced; so that in 2018 public expenditures per capita in in Italy is 120 euros, as compared to 160 in Spain and 380 in France and Germany. But the reduction was much larger in the Centre-South, due to a long series of discretionary and highly controversial changes in the allocation mechanisms. As a consequence, some university, all located in the cities of the "Cross" were able to preserve their financing; while public expenditure was reduced up to 25% in real terms in a few years in most of the universities located in the weakest regions of the North (such as Liguria and Friuli Venezia Giulia), in the Centre and the Mezzogiorno. University in Sicily were particularly hit.

In more recent years there has been a growing political pressure towards concentrating public expenditures in more developed areas of the country. For example, in 2017 a political campaign was launched in the three relatively stronger regions, namely Lombardy, Veneto and Emilia-Romagna, to achieve a "differentiated" regional autonomy; it means giving to those regions more extensive powers to implement public policies than in other areas of the country, together with an increase— compared to the rest of the country—in public financial resources. On the contrary, the political support for regional development policies, aimed at reducing both inter-regional and intra-regional disparities, seems to be at a quite low level. At the same time, policies aimed at facing the problems of "internal areas" are receiving a quite limited interest and support in the political arena.

Summing up, the issues of territorial development and territorial and regional disparities in contemporary Italy should deserve the utmost attention, given the national and local trends that have been sketched and the political trends that have been mentioned.

References

Coniglio ND (2019) Aiutateci a casa nostra. Perché l'Italia ha bisogno di immigrati. Laterza, Rome, Bari

Banca d'Italia (2019) L'economia delle regioni italiane. Rome

De Rossi A (2018) (Ed) Riabitare l'Italia. Le aree interne fra abbandoni e riconquiste. Donzelli, Rome

Fondazione Agnelli (2020) Rapporto sull'edilizia scolastica. Laterza, Rome-Bari

Golini A (2019) Italiani poca gente. Il paese ai tempi del malessere demografico. Luiss University Press, Rome

Iuzzolino G, Pellegrini G, Viesti G (2013) Regional convergence. In: Toniolo G (eds) The Oxford handbook of the Italian economy since unification. Oxford University Press, Oxford, pp 69–107

Joint research centre of the European Commission, Eurofound (2019) European Jobs Monitor 2019. Shifts in the employment structure at regional level. European Jobs Monitor series, Publications Office of the EU, Luxembourg

Manduca RA (2019) The contribution of national income divergence to regional economic divergence. Soc Forces 98(2):622–648. http://doi.org/10.1093/sf/soz013

OECD (2018) Productivity and jobs in a globalized world. (How) can all Regions benefit? OECD, Paris

Pugliese E (2018) Quelli che se ne vanno. La nuova emigrazione italiana. Il Mulino, Bologna

Svimez (2019) Rapporto Svimez sulla società e l'economia del Mezzogiorno. Il Mulino, Bologna

Viesti G (2016) Disparità regionali e politiche territoriali in Italia nel nuovo secolo. In: Mazzola F, Nisticò R (Eds) Le regioni europee. Politiche per la coesione e strategie per la competitività. FrancoAngeli, Milano

Viesti G (2018) La laurea negata. Le politiche contro l'istruzione universitaria. Laterza, Rome-Bari

Viesti G (2019b) Il trasporto ferroviario in Italia nel XXI secolo. Un paese sempre più diseguale. In Economia e Politica, December 2 https://www.economiaepolitica.it/_pdfs/pdf-10999.pdf

Viesti G (2019a) Qualche riflessione sulla nuova geografia economica europea. Meridiana 94(1): 137–164. https://doi.org/10.23744/2244

Viesti G (2019c) Verso la secessione dei ricchi? Autonomia regionali e unità nazionale. Laterza, Rome-Bari

Chapter 6
Socio-spatial Inequalities in Urban Peripheries: The Case of Italy

Gabriele Pasqui

6.1 Introduction

The theme of inequalities in Italian urban peripheral areas should not be separated from a more general reflection on the territorial gaps that have characterised Italy since its establishment as a national state, and that are more and more important now after the pandemic emergency.

The aim of this chapter is therefore to discuss the problem of inequalities in Italian urban contexts against the backdrop of a more general reflection on Italy's territorial fragility, the latter deeply influencing the dynamics of economic and social development.

The reflection focuses on the possible interpretations of the relationships between inequalities and peripheries from the perspective of space, namely the role of territorial variables in the determination of vicious circles that have generated a growing process of polarization between areas, neighbourhoods and municipalities within large and medium Italian urban areas.

In Sect. 6.2, we will consider territorial inequalities at national level in relation to the issue of internal gaps in urban and metropolitan contexts. Section 6.3 will present the issue of measuring and delimiting "critical" peripheral areas. Section 6.4 will discuss some elements for correcting a strictly quantitative approach to the identification of peripheral areas and its consequences for public policies. Section 6.5 will offer suggestions for an integrated approach to the theme of inequalities in the peripheries, with an in-depth analysis of employment, housing and education policies.

G. Pasqui (✉)
Department of Architecture and Urban Studies, Politecnico di Milano, Milan, Italy
e-mail: gabriele.pasqui@polimi.it

© The Author(s), under exclusive license to Springer Nature Switzerland AG 2020
A. Balducci et al. (eds.), *Risk and Resilience*, SpringerBriefs in Applied
Sciences and Technology, https://doi.org/10.1007/978-3-030-56067-6_6

6.2   Urban Peripheries and Territorial Inequalities

Italy's development dynamics have always had a complex territorial articulation, one that does not reflect the gap between central-northern and southern regions, and which has not narrowed over the past twenty years, indeed it has widened in some respects. More generally, Italian development, in its peculiar forms embodied by "territorial capitalisms", has expressed the same global dynamics affecting all European countries in the last twenty years, which have been articulated on a regional, sub-regional and urban scale (Triglia and Burroni 2009).

This territorial variety of forms and models of territorial development, which represented a significant strength of the country for a long time, became even more complex during the long crisis that began as the first decade of the 2000s drew to a close and also involved the dynamics of demographic contraction affecting Italy. In point of fact, the effects of the crisis, together with profound social and demographic processes, redesigned the outright geography of territorial development models, with significant effects on the overall performance of the Italian economy.

However, we cannot forget that productivity performance, development processes, and the growth of inequalities are part of a more general European dynamic. The problems of regions and cities in crisis, in the context of globalization processes, constitute a central element of understanding the cultural and political erosion of the consensus towards the European Union.

A reflection on the problems and perspectives of Italian development from the territorial standpoint can only start from the observation that despite a season of "new programming" promoted by Prime Minister Carlo Azelio Ciampi in the 1990s, launching well-designed policies, in the last thirty years regional, infra regional and infra-urban gaps have not been closed. In this context, in the last ten years many "local capitalisms" (especially located in industrial districts, medium-size cities, diffused industrialization areas), which have long characterized our country's development model by putting to work a set of rooted and localized resources of material and immaterial nature, seem to have exhausted their driving force.

This is not just an Italian problem: the challenges of European development are mainly territorial. In general, the European economy (and society) is characterized today by levels of inequality (capital, assets, income, access to services, training) between national states, regions, urban areas, dynamic and marginal areas, that are greater than twenty years ago. Although regional disparities have narrowed over the past three years, inter- and intra-regional divergences constitute one of the strongest threats to defining a new development path for the European economy (European Commission 2017).

Some recent studies explore the link between productivity deficits and territorial performance (education and training levels, fundamental public services to businesses and persons, efficiency and effectiveness of local public administrations, tangible and intangible infrastructures). These differences in performance generated highly and increasingly differentiated development patterns, outlining a new European economic geography (Iammarino et al. 2017). The fundamental features of this geography also depend on the growth of inequality between people, which has important consequences on the inequality between neighbourhoods, cities and regions. In turn, this inequality is connected to the profound revolution in economic activities, due to the spread of new technologies and to the consequent new division of labour between countries and within each country, with a growing role played by urban economies.

As copious interdisciplinary literature shows, qualified human capital and the production of innovation are concentrated in the cities. It is the urban space that defines the connections between flows of technologies, people, and goods on a global scale. On the other hand, the phenomena of social vulnerability, the poverty traps, the processes of loss of social cohesion, the conflicts between groups, populations and social classes are also concentrated in the urban space.

This set of phenomena leads us to doubt theories of regional and urban convergence, upon which many territorial development policies have been based over the past fifty years. Furthermore, three strictly European factors (the enlargement of the European Union to the East, the sovereign debt crisis, and the demographic transition) have reshaped the continent's economic geography, articulating it on new east–west axes as well as on the more traditional north–south axis. Along these axes, Italy and other countries have experienced a sharp slowdown in productivity and a relative deterioration in economic performance compared to European averages in the new century (Viesti 2019).

These processes have also had very significant political consequences, which became evident in the last European Parliament elections and in many domestic political systems, including in Italy, helping to bring the issues of the territorial dimension of development to the centre of the agenda. It is not just about the affirmation of so-called "sovereign" political forces; more generally, some scholars have spoken of a "revenge of the places that do not count" (Rodriguez Pose 2017), which risks undermining the actual European perspective.

In this general context, the Italian case may be considered an interesting example. Negative dynamics of Italian productivity, in their consequences for urban economies, must necessarily be correlated to the performance of that set of goods and services, public and private, which literature has defined "territorial capital" (Fratesi and Perruca 2017). For this reason, territorial development policies sensitive to places can play an important role in rescuing Italy and its urban areas from the overall risk situation.

Recent literature has clearly identified some features of the structural deficit of Italy's territorial capitalisms. These traits are also extremely important for discussing inequalities and gaps within urban areas. Nonetheless, it is important to recall some features of Italy's territorial dynamics of development before we start.

The first is connected to the identification and mapping of the territorial articulation of critical situations. This articulation, recently studied by literature that renewed the economic and social analysis of territorial development, allows us to identify new ruptures. These ruptures don't include only the traditional divide of north and south, or between the so called "three Italies" (Dunford 2008), but also "horizontal segmentations", within macro-regional contexts, which redefine a new geography of "full" and "empty" regions (see Cersosimo et al. 2018). The analysis of this geography does not undermine structural elements of the traditional distinctions, starting from the divide between the south (and parts of the centre regions) and the rest of the country. Rather, it allows identification of a map of Italy in contraction, which pinpoints not only the inland districts (highlands and abandoned villages) but also shows rural production areas in depopulation, district urbanization in decline, coastal territories impoverished and consumed by mass tourism, suburbs and fragile urban interstices (see Lanzani and Curci 2018).

Differences and ruptures are therefore identifiable within urban areas that in their own turn are undergoing profound change, where inequalities between urban sectors and between neighbourhoods have grown substantially, but also between urban areas and their hinterland (with a real risk of "divorce", for example, between Milan and its metropolitan region).

This territorial articulation obviously depends on multiple technological, demographic, social, and cultural factors, but is certainly related to the differentiation in the collective production capacity of public, material and immaterial goods, and infrastructures. It is not just about the lack or inefficiency of essential services in some areas of the country (accessibility through public transport, primary education, basic health and social services). It is a more general deficit of fundamental goods and services both for the daily life of citizens, and for the efficiency and competitiveness of businesses.

This crisis in the production of essential territorial public goods and services represents one of the most important factors in interpreting the difficulties of the Italian economy. It is also closely related to the disappointing performance of public administrations, both from the point of view of regulation and the production of goods and services, as well as in terms of skills and resources for the design and implementation of urban and territorial policies, in the absence of a clear national policy agenda.

Territorial capitalisms in Italy have always fed on a set of factors capable of strengthening competitiveness and productivity. These factors have been linked to the relationships between businesses and people, to the characteristics of local societies and their subcultures, to forms of social and political intermediation. However, in the face of the aggressive technological, organizational and cognitive processes connected to globalization, in order to survive and grow, these capitalisms need to take root in territorial contexts in which fundamental goods and services are produced not only with efficacy and efficiency, but also available to all and in a perspective of new citizenship.

This depends among other things on the change in the economic and technological forms within which the value chains are structured. For example, the change in urban economies downstream of the crisis that began in 2008–9 is not only cyclical. The real estate stagnation in Italy, with the partial exception of the Milan and Bologna areas, and a few other cities, leads us to reflect carefully on the breaking of the long-term link between urban development and growth, which had characterized Italian development from the mid-twentieth century. New urban economies (linked to tourism, for example, but also to the production of high and low value-added services) have been accompanied by significant contraction both in terms of population and of real estate investments. The challenge for urban economies is the definition of development paths decoupled from settlement growth and consumption of non-urbanized land, being oriented rather towards promoting processes of regeneration, adaptive reuse, and requalification of extant heritage.

This change in the medium-term dynamics of the urban market acquires an even stronger condition if we consider the issue of environmental sustainability. The crisis of consolidated and development models is rooted in the costs of failing to prevent the possible effects of the various dimensions of territorial fragilities, from seismic to hydrogeological risk, maintenance of material welfare, and the theme of adaptive reuse and recycling of built heritage.

The governance of global sustainable regeneration processes, in the face of an ecological challenge and climate change, whose relevance could not be overestimated especially after COVID emergency, has essential implications on the strategy and design of territorial policies for development and cohesion, with particular reference to urban areas.

Of course, not all problems can be addressed through territorial and local policies: economic policies on a European and national scale remain crucial. However, due to its nature and history, Italian development has to be articulated territorially: for this reason, public action cannot ignore the relevance of territorial policies for the identification of strategies, programs and projects capable of enabling and promoting a credible dynamic of sustainable development.

This apparently obvious statement is not at the centre of policymaker reflections and actions. On the contrary, the lack of a national strategy for territorial development—in particular for cities—represents one of the greatest cultural rather than political weak points of government action at different levels (Urban@it 2017, 2018, 2020).

If this diagnosis in a nutshell is sufficiently robust, the priority that comes to the fore is of a cultural and methodological nature, addressing the problems of Italian development in the urban context also, both from a low productivity and inequality growth perspective as far as territories are concerned.

This means investigating in depth growing territorial inequalities, with particular reference to those within the urban areas, in order to build actions capable of recognizing critical issues, but also territorial specificity and resources. This is certainly not an unprecedented perspective: it was precisely Italy, from the latter half of the 1970s, that built up cognitive conditions for attention to the place-based dimension of public development policies. The same approach adopted by the European Commission from the 1990s and summarized effectively but also critically by the *Barca Report* (Barca 2009), was nurtured by the awareness that sustainable development needs strong territorial infrastructure. In this perspective, the effectiveness of development policies depends on the ability to promote "development strategies rooted in places". These strategies should be capable of designing and implementing—primarily in the country's fragile and peripheral areas—policies grown by the engagement of inhabitants to combine the improvement of fundamental services with the creation of opportunities for a fair and sustainable use of new technologies.

Moreover these strategies should be able to identify development patterns capable of putting underutilized human and physical resources and capital to work.

6.3   Which Peripheries? Which Inequalities?

In the light of the arguments above, we can now focus on the theme of urban peripheries.

In the Inequality and Diversity Forum (Forum Disuguaglianze e Diversità) report entitled "15 Proposals for Social Justice", the theme of urban suburbs is addressed within the broader framework of territorial inequalities (Forum Disuguaglianze e Diversità 2019). As widely documented by the report and more generally by the materials produced and collected by the Forum, territorial inequalities have increased in recent decades throughout the Western World, particularly Italy.

Reflecting on the relationship between peripheral conditions and the growth of inequalities means first of all placing the theme in the scenario of the expanding divergence between regions and areas of the country. As the report says: "On a territorial level, the situation in Italy is particularly serious: due to the size of the gaps between regions (for example, the average monthly disposable income in Lombardy is 69% higher than that of Calabria), and because all Italian regions have lost ground compared to the rest of Europe. For example, between 2003 and 2017, Lombardy went from 28th to 52nd place in the ranking of European regions in terms of per capita GDP; Emilia-Romagna from 45th to 72nd." (Forum Disuguaglianze e Diversità 2019, 19).

If urban peripheral areas are places characterized by growing inequalities, what are we actually saying about when we talk about "peripheries"? As Agostino Petrillo (2018) has shown, the notion of "periphery", and the delimitation of peripheral areas, constitutes a complex research problem. From this point of view, recent attempts made in our country to determine the minimum territorial investigation units and to characterize in a statistical sense the territorial forms of socio-economic inequality in urban areas make it possible to highlight some relevant analytical and interpretative problems.

We are unable to discuss these attempts in detail here so we will briefly analyse two works of great importance, carried out respectively by the Department for Cohesion Policies (Dipartimento per politiche di coesione. Nucleo di valutazione e analisi per la programmazione) and by ISTAT, the Italian institute of statistics, both published in 2017.

The Department has drafted poverty maps by identifying sub-areas or districts of concentrated distress in each of the fourteen Italian metropolitan areas recognized by the legislator. In this way, it is possible to classify and characterize their features and provide an operational tool in support of urban policies and local planning (Dipartimento per politiche di coesione 2017).

ISTAT, on the basis of a request that emerged from the Parliamentary Commission of Inquiry into the conditions of safety and state of degradation of cities and their suburbs (Camera dei Deputati 2018), produced and calculated an indicator of social and material vulnerability for the fourteen identified municipal districts. The indicator has been created by combining different indicators and aims to produce a unitary and georeferenced representation of inequalities in the urban context using cartograms (ISTAT 2017).

Despite the significant differences between the two works, the efforts of the two reports aimed first of all to identify the minimum territorial unit of reference, secondly at producing concise, multidimensional indicators able to identify and systematically represent the territorial morphology of the distress.

The Department adopted the neighbourhood as a reference unit, even though the authors of the report openly acknowledge there is no clear theoretical or statistical definition of the term. From this perspective, use of a minimal territorial proxy of the neighbourhood concept based on ISTAT census areas constitutes a problem which it is interesting to consider. Based on these and other methodological choices, poverty maps build a classification of neighbourhoods by levels and types of hardship.

The ISTAT report, on the other hand, assumes different historical and functional—not always comparable—subsets, reconstructing concise values for these areas on the basis of a multidimensional battery of indicators calculated by referring to different statistical sources.

These are two extremely important approaches as they allow us to reason on the basis of accurate and comparable data and to define with precision the different variables involved. Furthermore, both works assume some guiding principles that also reveal an implicit theory of the nature of inequalities concentrated spatially in urban areas. In brief, the guiding principles seem to be:

- financial poverty (absolute or relative) is strictly connected to housing poverty and therefore plays a central role in determining the conditions of spatial concentration of families and individuals who suffer conditions of economic inequality;
- inequalities concentrated only in some areas of the cities, on the other hand, are a multidimensional phenomenon, in which social vulnerability, connected to precarious employment conditions, level and quality of education and more generally of training, accessibility of services, and mobility potential play a role;
- building degradation and related fragility of housing conditions provide a good approximation of the environmental variables characterizing the districts and areas of worst deprivation.

The Dipartimento per le Politiche di Coesione and ISTAT reports focus on non-comparable phenomena (poverty for the former; vulnerability for the latter), non-uniform analysis units and dissimilar contexts, but nonetheless offer a very rich basis for reflection and reasoning on spatial types of inequalities in Italian peripheries.

The main results of the two surveys show an articulated geography of the suburbs in large Italian urban areas, but also the persistence of critical areas, concentrated in the larger public housing neighbourhoods built in the twentieth century and in some degraded historical districts.

6.4   History and Places Matter

Now I would like to propose some specifications to the works synthetically presented from a different perspective, which takes places and history as privileged points of observation.

The first integration relates to the strong historical dimension of the dynamics of inequality. The Italian cities, over the decades, have repeatedly redesigned the geography of their peripheries, in relation to demographic, social and territorial processes that should be analysed in depth to understand the historical dimension of the phenomena.

European metropolitan peripheries have long been places characterized by distance and separation (physical and symbolic) from the centre, but also contexts in which highly innovative planning, design and policies were experimented (Secchi 2005). Over the last quarter century the geography of the suburbs has changed radically with respect to the breakdown of the traditional urbanization processes. ,

There are some features of this change that should, of course, be tailored specifically to each different area of the country and read in tandem with the material and symbolic evolution urban forms (Balducci et al. 2017). Moreover, these processes are deeply connected to the history of the structural crisis of material and immaterial welfare.

The first trait of these dynamics is the very strong connection between demographic change and peripheral condition. The peripheries are where the two processes that characterize Italian demographic dynamics are concentrated: an aging Italian population (often the result of the continuous presence of families in public housing districts) and the significant presence of immigrant population (legal and illegal). The demographic dynamics in the peripheral areas are too important to be ignored. Many of the historic suburbs of Italian cities began life as unitary settlements designed to house urbanized worker families, often characterized by stable working conditions and expectations of an improved economic and social status. Today it is precisely the joint presence of an elderly Italian population and a young, foreign population suffering precarious employment conditions, that makes social cohesion more of a challenge and increases the perception of insecurity.

The second aspect is the spatial distribution of difficult conditions and consequently inequalities. Over time, many peripheral areas have been engulfed by the city and subject to significant phenomena of regeneration or degradation and downgrading. Neighbourhoods considered examples of peripheral condition and social vulnerability today are urban districts whose quality of life and social cohesion are more than satisfactory. On the other hand, there are neighbourhoods, but also parts of a neighbourhoods, in which an accumulation of causes have led to degradation phenomena. Furthermore, the places of greatest deprivation today are often close to the city centre, concentrated in territorial pockets (a building, a few streets, the old town district, an area close to the railway station ...) which are often difficult to detect on a statistical level and cannot, in any case, be ascribed to traditional neighbourhood geography.

Third: if the definition of periphery in the twentieth century often coincided with a concentration of public housing located there, subject to phenomena of social and material degradation, today some of the most challenging situations concern neighbourhoods or urban sections where private ownership prevails. In these areas the replacement of the resident population and the urban market crisis have generated housing and social vulnerability that are quite often an explosive mix.

The fourth trait, closely connected to the other three, concerns the increasingly strong connection between peripheries and the perception of insecurity (Palidda 2016). The growth of inequalities apparently has no direct connection with the theme of the growing perception of insecurity in urban contexts but increasingly the story of peripheries coincides with a narrative of insecurity, in turn rooted in demographic (aging native population and increasing non-native population) and socio-economic (especially in terms of job insecurity) processes.

The fifth and final aspect concerns the relevance of the narrative and symbolic dimension of the processes on the basis of which conditions of periphery stigmatization and self-stigmatization are determined. Little will be understood of the suburbs, and the connection between these spaces and the concentration of inequalities if we do also address the narratives, the stories, the way in which the images settle down over time, working on expectations and on aspirations (Bourdieu 1993; Appadurai 2004). Peripheries are places that their inhabitants often perceive as an inescapable destiny, perhaps even as areas of expression of a profound social and cultural otherness. This perception of living in the suburbs as a fate sometimes becomes a trap that is more insidious than financial or housing poverty.

6.5 An Example: Milan

Unfortunately, it is not possible to show empirical materials to offer evidence of this reading of the disparities that have occurred in recent years in the spatial and social forms of peripheral areas. Nonetheless, an observation of the case of Milan confirms the elements we have just identified (Pasqui 2018).

For Milan, the peripheral conditions are not only geographical (distance from the centre), nor do they coincide solely with public housing neighbourhoods. Indeed, some of the large public districts are now liveable places, with high green standards and services and good urban quality. Even neighbourhoods traditionally suffering very strong stigmatization, now vaunt by far better conditions than in the past thanks to efficient public policies and the commitment of determined, competent players. A case is Ponte Lambro district, south-east of Milan municipality. There are also enclaves of hardship within amply regenerated neighbourhoods (this is the case of Via Bolla, in the Gallaratese district), or pockets very close to the centre, in which conditions of insecurity and social fragility are concentrated (Via Gola, Viale Bligny). Today it is often privately-owned real estate in Milan to find itself in difficult conditions of material degradation and social hardship. Often these areas are not located in the region's capital city: to give just two examples, the Crocetta district in Cinisello Balsamo or the Satellite district in Pioltello are situations that present a challenge for public policies.

Therefore, new geographies of Milanese peripheries are often concentrated in municipalities other than that of the capital, which implies that the dynamics of "peripheralization", in the case of Milan, have a strong connection with processes of re-concentration. To understand the phenomenon, we must think that Milan is many different things, and that in all of them there may be of potential growth processes for social and spatial inequalities.

In a phase in which narratives and public imagination give a dynamic, positive representation of the city, with the expression "Milano" we embrace very different realities, although closely intertwined with each other.

Milan is first and foremost the central city, the capital city of this region, covering a small area (181 sq. km, compared to Rome's 1,285 sq. km), still confined in its historical boundaries. A very strong narrative has grown around this city in recent years, also deeply rooted in the city's long tradition and history, but fuelled by a change in expectations also perceived by international investors. For some observers, this Milan, before COVID, was experiencing a "magical" moment, of which Expo 2015 represented the symbolic, inaugural event: a dynamic, welcoming city, with its universities (just under 200,000 students registered, almost 40,000 staff including faculty and technical and administrative workers) and its excellence (finance, fashion and design, economies of culture and communication, health with related technologies). A city with a strong European vocation, attractive to foreign students, as well as to temporary users of the many successful events, regular users of the city, and also migrants and—more recently—tourists (more than ten million in the metropolitan city in 2019), especially from abroad.

Over the years, this Milan has changed primarily in an incremental way, through molecular processes and mechanisms that involved the social mobilization of families and businesses, rather than as a result of comprehensive planning. However, in this central heart of the urban region, important domestic and international investments have consolidated in recent years. The Qatar sovereign fund acquired the entire Porta Nuova project, for a market value that observers estimate to be in the region of two billion euros, but also operations such as the opening of the Apple Store in Piazza Liberty, a stone's-throw from Duomo, show how Milan has returned confidently to the map of both large international financial investors and globalized multinationals.

It is here that the processes of social innovation find fertile ground, not least because of the permeability of urban spaces. However, even in this part of the city there are derelict areas, places of deprivation and degradation (Via Gola, Via Bligny, areas close to the central railway station, are just a few examples), although Milan's urban heart really seems to be at the centre of a small renaissance, also fuelled by the repurposing of long abandoned spaces.

Milan, however, is not only this. It is also the changing city that extends, with variable geometry, between the municipal borders and the conurbation of the municipalities of the first and second belts. It is in this city that some of the most important transformations have taken place or could still occur, and it is here that the contrast between economic and social dynamism and new forms of inequality and fragility appears stronger.

In this intermediate city there are large real estate operations: Cascina Merlata residential development; MIND project on the post-Expo area, led by the Australian developer Landlease; the investments by mega retail chains in Arese, Cinisello Balsamo, Segrate; the relocation of important hospitals in Ronchetto sul Naviglio or in the abandoned industrial areas of Sesto San Giovanni. In the metropolitan area are located also the old and new suburbs, characterized by social and economic fragility and the cultures of "resentment".

In this intermediate city, which goes beyond the municipal borders, the places of social crisis have changed compared to the geographies of the "public" city and the large districts to which we had become accustomed. The most worrying social and demographic dynamics no longer concern only the enclaves of Corvetto, Gratosoglio, Quarto Oggiaro or San Siro, but also the private owned neighbourhoods of Crocetta in Cinisello Balsamo or the Satellite in Pioltello.

In this "middle Milan", therefore, differentiated and sometimes contradictory dynamics and processes overlap, in which social composition and political cultures are unreadable. It is no coincidence that in the last general elections (March 2018), the divergence between the Left and Centre-Left parties and the popular classes was far more evident here.

Milan is also a large urban region that important studies (Balducci et al. 2017) define post-metropolitan. It is an integrated urbanized area at the centre of Lombardy's production platform, where the balance and complementarities between the capital city and the region have long represented an extraordinary competitive advantage. This Milan extends between the foothills (Novara, Varese, Como, Bergamo) and the irrigated plain (Pavia, Lodi, Piacenza), and is structured on a complex context of interrelationships, long and short networks, economic relations between supply chains and territorial clusters. Precisely with respect to these networks in recent years it has been possible to pinpoint a risk of divergence between the city and the region, between Milan and its hinterland.

This large region, in turn, is part of an enlarged urban context (a mega-city region, as Peter Hall would say), which extends (at least) from Turin to Venice, in a logic of complementarity and competition in which infrastructure programs and functional connections play an essential role. In this mega-city region, the peripheries are often rural areas, medium- and small-size cities excluded from metropolitan dynamics.

Finally, Milan is the gateway to global flows, a connector city located within international networks that go beyond geographical proximity and mobilize significant financial investments, but also flows of qualified human capital (Bolocan Goldstein 2017).

Each of the five Milan shows potential signs of polarization, of socio-spatial differentiation. Only by considering these processes together is it possible to understand the need for specific interventions, for very different places and areas.

6.6 Policy Guidelines

In the context we seek to spotlight, it becomes important to recognize the central nodes of the territorial dynamics of inequality, and in that scenario also focus on possible policies, projects and actions aimed at increasing social and spatial justice.

What path do we take if we want to deal with the different dimensions of inequality in urban peripheries and design effective public policies to deal with these problems? The Inequality and Diversity Forum report proposes to revive a place-based approach for peripheral and inner areas (Barca 2009), starting with "the needs and aspirations of people in places" (Forum diseguaglianze e diversità 2019, p. 97). By putting in place such an approach, suspending sectoral policies and reassembling them as a territorial development project, it would be a matter of "designing and implementing [...] development strategies in the peripheries aimed at places that learn lesion in direction and method from the National Strategy for inner areas; strategies that combine the improvement of fundamental services with the creation of opportunities for a fair and sustainable use of new technologies with a strong engagement of the inhabitants" (ibidem, p. 14).

The Forum's perspective is shared by many and assumes that in relation to territorial inequalities, the classic response of a redistributive nature is not sufficient (Franzini et al. 2016). We must rethink and evaluate carefully the season of place-based urban policies, which is behind us today, to effectively recalibrate interventions in the current critical phase of European and national urban policies (Urban@it 2017, 2018, 2020).

In order to support an urban policy capable of reconciling innovation and inclusion as far as possible, it is essential to put in place a public action focused on the regeneration of urban peripheries.

Urban regeneration programs and projects for the most degraded peripheral areas in urban contexts are not just a question of social policy. On the one hand, they enable mobilization of resources (starting from human capital) and creation of new (and good) jobs. On the other hand, they reduce the risks associated with the (economic and social) costs of inequality (Roberts et al. 2017).

Unlike what happened in recent years with some national programs for the peripheries (from the "Piano Città" to the "Piano Periferie"), it is not a matter of emergency responses. It is a matter of building—patiently and in close relationship with regional and local authorities—an integrated urban regeneration program of the peripheral areas developed through participatory and place-based logic. To build new policy tools effectively for reducing urban inequalities, we need to spark three engines: employment, home, school.

The first engine is the construction of active employment and training policies on an appropriate territorial scale, capable of intercepting above all the most fragile sections of peripheral populations, namely those who are excluded from the labour market or whose working conditions are extremely precarious. Various experiments have been conducted on this front, which have so far been not very effective, because they often cannot intercept the truly disadvantaged sections of the population. Place-based projects capable of generating proper employment in peripheral areas of our cities would first of all need to recognize a central objective, which is the maintenance of equipment and material welfare (public housing assets, sports facilities, schools, parks and gardens, etc.). These could be a relevant public investment, in which the direct participation of the inhabitants in caring for the places where they live and spend their time also becomes an opportunity for creating better-qualified and less precarious work (also through suitable training courses).

The second engine is the investment in housing. Housing hardship and poverty today are perhaps the most challenging hurdle of life in the peripheries. Housing policies on a national scale are also fundamental.

However, there is a strongly rooted dimension of housing policies suffering actual legislative conditions and scarce resources but which would require identification of managerial and administrative innovations, capable of improving housing conditions for the least affluent and of creating significant experiences of mobilizing underutilized local resources around living practices.

Finally, the third engine is patient, in-depth work on schools as powerful places of integration. Schools in the suburbs are remarkable sounding boards for social innovation, in which voluntary mobilization and social commitment find a privileged environment. This overview of schools must not overlook the fact that they are also places of conflict, where many compete for scarce resources. Recognizing conflicts and tensions means quashing any nostalgia for community relations and closeness that cannot be reproduced in the context of pluralization of life forms. On the other hand, there is no denying that schools are potential places for fertile interaction between different people, in which it may be possible to work towards achieving the tricky balance between universalistic demands and the proliferation of variety. For this reason, working in and on peripheral schools means understanding not only how socio-spatial segregation processes are investing in school systems in new forms, including the field of public education, but also how schools can be understood as effective experimental welfare platforms. To move in these interconnected directions, investments and policies for schools and in particular for school building heritage must acquire the characteristics of integrated regeneration and development projects. Projects of this nature would have the merit of activating administrative action in a joint and coordinated form and would take on the role of integrated place-based interventions in the context of cohesion policies. In analogy with these policies, "school contracts" could be developed resembling "neighbourhood contracts", which could also be implemented in the form of competitive tenders, on the condition of favouring access to financing also by schools that pay in situations of material deprivation and poor maintenance or which are located in particularly poor urban contexts.

This three-engine approach might make it possible to build a real national agenda for suburbs, which would take charge of tackling the problem at territorial level. In brief, employment, housing and education are the three critical fields necessary to guarantee the effectiveness of actions and policies. It is precisely on this terrain that the game of peripheral policies as actions for social and spatial justice should be played out today.

References

Appadurai A (2004) The capacity to aspire: culture and the terms of recognition. In: Rao Y, Walton M (eds) Culture and public action. Stanford University Press, Stanford CA, pp 59–84

Balducci A, Fedeli V, Curci F (2017) (eds) Post-metropolitan territories: looking for a New Urbanity. Routledge, London

Barca F (2009) An agenda for the reform of cohesion policy a place-based approach to meeting European Union challenges and expectations European Commission. Independent Report, Bruxelles

Bolocan Goldstein M (2017) Geografie del Nord. Maggioli, Santarcangelo di Romagna

Bourdieu P (1993) La misère du monde Seuil. Seuil, Paris

Camera dei Deputati (2018) Relazione della Commissione parlamentare di inchiesta sulle condizioni di sicurezza e sullo stato di degrado delle città e delle loro periferie. Roma

Cersosimo D, Ferrara AR, Nisticò R (2018) L'Italia dei pieni e dei vuoti. In: De Rossi A (eds) Riabitare l'Italia. Le aree interne tra abbandoni e riconquiste. Donzelli, Roma, pp 21–50

Dipartimento per le Politiche di Coesione (2017) Poverty Maps. Analisi territoriale del disagio socio-economico delle aree urbane. Roma

Dunford M (2008) After the three Italies the (internally differentiated) North-South divide: analysing regional and industrial trajectories. Annales Dés Geographie 6(664):85–114. https://doi.org/10.3917/ag.664.0085

European Commission (2017) My region, my Europe, our future: the seventh report on economic. Social and Territorial Cohesion, Brussels

Forum Diversità e Diseguaglianze (2019) 15 proposte per la giustizia sociale. Roma

Franzini M, Granaglia E, Raitano M (2016) Extreme Inequalities in Contemporary Capitalism. Springer Verlag, Berlin

Fratesi U, Perruca G (2017) Territorial capital and the resilience of European regions. Ann Reg Sci 60(2):1–24. https://doi.org/10.1007/s00168-017-0828-3

Iammarino S, Rodríguez-Pose A, Storper M (2017) Why regional development matters for Europe's economic future. European Commission, Working Papers, WP 07/2017, Bruxelles

ISTAT (2017) Materiali per la commissione parlamentare di inchiesta sulle condizioni di sicurezza e sullo stato di degrado delle città e delle loro periferie. Roma

Lanzani A, Curci F (2018) Le Italie in contrazione, tra crisi e opportunità. In: De Rossi A (ed) Riabitare l'Italia. Le aree interne tra abbandoni e riconquiste. Donzelli, Roma, pp 79–107

Palidda S (ed) (2016) Governance of security and ignored insecurities in contemporary Europe. Routledge, London, New York

Pasqui G (2018) Raccontare Milano. Politiche, progetti, immaginari. FrancoAngeli, Milano

Petrillo A (2018) La periferia nuova. Disuguaglianza, spazi, città. FrancoAngeli, Milano

Roberts P, Sykes H, Granger R (eds) (2017) Urban regeneration. A handbook. Sage, London

Rodríguez-Pose A (2017) The revenge of the place that don't matter (and what to do about it). Cambridge J Reg Econ Soc 11(1):189–209. https://doi.org/10.1093/cjres/rsx024

Secchi B (2005) La città del XX secolo. Laterza, Bari

Triglia C, Burroni L (2009) Italy: rise, decline and restructuring of a regionalized capitalism. Econ Soc 38(4):630–653. https://doi.org/10.1080/03085140903190367

Urban@it (2017) II Rapporto sulle città italiane. Le agende urbane delle città italiane. Il Mulino, Bologna

Urban@it (2018) III Rapporto sulle città italiane Mind the gap. Il distacco tra politiche e città. Il Mulino, Bologna

Urban@it (2020) V Rapporto sulle città italiane. Politiche urbane per le periferie. Il Mulino, Bologna
Viesti G (2019) Qualche riflessione sulla nuova geografia economica europea. Meridiana 94:137–
    164. https://doi.org/10.23744/2244

Chapter 7
Natural Risks Exposure and Hazard Avoidance Strategies: Learning from Vesuvius

Francesco Curci

7.1 Introduction

This chapter considers for the first time in English some essential issues dealt within the *Rapporto sulla promozione della sicurezza del patrimonio abitativo* (Report on the promotion of housing stock safety) drawn up by the Struttura di Missione Casa Italia ('Casa Italia' Task Force) under appointment of the Italian Presidency of the Council of Ministers (Casa Italia 2017). In particular this essay traces a part of the considerations contained in the fifth chapter of the report—edited by the author as a member of the task force—devoted to discussing the approaches and features of public policies aimed at containing or reducing residential exposure to natural risks in the Italian context.[1]

The conventional concept of risk in relation to human settlements depends on the hazard, exposure and vulnerability (Varnes 1984; UNDHA 1992). Therefore, one of the paramount items of evidence related to natural disasters is that the exponential growth of world population (doubled in the past 50 years) and its concentration in some urban areas has been increasing risk dramatically (UNDRR 2019). For this reason, even though difficult to fulfil, a more aware and intelligent location and distribution of buildings and population—chiefly residential buildings and resident population—should be a priority choice to reduce disaster risks. But this purpose is plagued by a series of dilemmas and trade-offs. Since different natural hazards may occur in different geographical contexts, exposure reduction is not always the most rational and feasible option. In Italy, for example, it has been realised that even though

---

[1] The Report is the result of reflection done by all the members and experts of the Casa Italia Task Force (*Struttura di Missione Casa Italia*) in charge between 2016 and 2018. It was established by the Italian Prime Minister after the earthquakes that hit Abruzzo, Lazio, Marche and Umbria Regions between the second half of 2016 and the beginning of 2017.

F. Curci (✉)
Department of Architecture and Urban Studies, Politecnico di Milano, Milano, Italy
e-mail: francesco.curci@polimi.it

it reduces risk exposure, the depopulation of mountains—which often coincide with the areas with the highest probability of earthquakes, landslides and floods—cannot represent a truly sustainable solution because of political, social, cultural and environmental reasons (Barca et al. 2014; Casa Italia 2017; De Rossi 2018). Another corroboration is that migrations from these 'inner areas' (*aree interne*) towards more densely populated can contribute to increase exposure to other natural and anthropogenic hazards. Consequently, exposure-based policies, which undoubtedly represent a strategic choice, strongly depend on different geographical and social variables. Generalised plans and actions aimed at reducing natural risk exposure are destined to failure because they do not take into account local specificities and geographical interdependencies. Accordingly, the Casa Italia Report does not propose an undifferentiated recipe but reflects upon a few emblematic cases with very problematic residential exposure (potential or existing) to different types of natural hazards. One of these cases (the Vesuvius volcano) is presented in this essay.

7.2 Why and When to Reduce Exposure as Priority Choice

In a milestone definition proposed 25 years ago by the U.S. Federal Emergency Management Agency we can find an explanation of the tight linkage existing between hazard and exposure as basic dimensions of risk mitigation strategies.[2] According to this definition *hazard mitigation* consists in a "sustained action taken to reduce or eliminate long-term risk to people and their property from hazards and their effects" (FEMA 1995, p. 3). Consequently, hazard mitigation strategies bring together *hazard avoidance* and *hazard reduction* strategies, but while the latter focus "on strengthening structures and providing safeguards to reduce the amount of damage caused by natural hazards [...], hazard avoidance strategies are designed to minimize the exposure to risks based on location" (NOAA Coastal Service Center 1999). Thus hazard avoidance strategies are aimed at minimizing exposure both in terms of number of lives (human, primarily) and number of assets (economic, social and cultural). These kinds of strategies actually embrace all those activities to be applied when it is not possible or sufficient to merely operate on hazard and vulnerability reduction, or on the increasing of communities' resilience. This is true when the potential hazard events are characterised at the same time by: (i) high destructiveness and potential mortality rates; (ii) very rapid occurrence; (iii) low temporal predictability; (iv) high spatial predictability. In line with the general approach of Casa Italia Task Force, this contribution does not make reference to emergency management approaches but focuses only on long-term prevention strategies based on residential exposure containment and reduction with particular attention to volcanic risk scenarios.

---

[2]This does not mean that there are no differences from an operational point of view. Indeed, while hazard mitigation measures are typically "structural", vulnerability and exposure reduction is mostly based on "non-structural" measures (UNDSR 2009; Pesaro et al. 2018; Menoni 2019). Nevertheless some measures aimed at reducing physical vulnerability and minimizing exposed settlements and infrastructures can be structural as well (Esteban et al. 2011).

7.3 Hazard Avoidance Approaches

Depending on the specific situations, both for individuals and organisations, hazard avoidance can assume different meanings based on how much: exposure is permanent or temporary, connected to work or leisure activities, subject to active or passive public policies; risk is perceived as probable or improbable; risk perception is tied to local myths and beliefs (religious or otherwise).

In general terms hazard avoidance strategies can be implemented in two ways: *preventing development* "through direct land management of public lands or through prohibitions on altering critical habitat"; or *managing development* "through retreat, zoning and subdivision regulations, land use plans, land acquisition, economic incentives, location of capital facilities, information dissemination, education and training" (FEMA 1995). Both these development behaviours are related to two main and complementary ways to reduce exposure. The first way consists in acting on *potential exposure* thus in inhibiting the settlement of new population and activities in high-level hazard areas; the second consists in acting on *existing exposure* thus relocating the already settled population and activities in more safe or less hazardous areas, removing or not the existing buildings (Table 7.1).

In the case of potential exposure reduction purposes, the main approaches are of the regulatory, disincentive and educational type. Through the combination of these three approaches it is possible to prevent the construction of new residential buildings in proven and declaredly hazardous places, but also the conversion into housing or accommodation facilities of buildings with other original functions—including the inhibition of abandoned buildings restoration. Combining the three approaches is also possible to encourage the transfer elsewhere of any building rights not yet exercised.

The *regulatory approach* is the most widespread in the world. It is linked to the regulation of land use (often through to the establishment of new protected areas). This type of approach, characterised by recourse to zoning and urban planning rules as a prerequisite, is based on the prohibition or limitation of building rights and permits for definite hazardous portions of territory. The importance of land use plans

Table 7.1 Hazard-avoidance tools according to exposure types and development behaviours

| | | Development | |
		Prohibited	Managed
Exposure	Potential	Building bans; Streams buffers and setbacks (through ordinances or codes)	Land use plans; Zoning and subdivision rules; Land acquisitions; Location of capital facilities
	Existing	Permanent evacuation orders; Building removal orders	Retreat plans and regulations; Relocation incentives; Building removal plans or guidelines

Source Produced by the author based on FEMA (1995)

in order to reduce risk in hazardous areas has been empirically demonstrated for many years. Some studies have given systematic evidence of the benefits deriving from the consideration land use plans as a vehicle for limiting development of areas at risk from natural hazards (Burby and Dalton 1994; Burby 1998).

The *disincentive approach* is based on the discouragement (bureaucratic, financial, fiscal) of building or inhabiting options in particularly hazardous areas in which there are no building or inhabiting bans.

The *educational approach*, that is in some ways analogous to the second but with a different nature, is based on making citizens aware of the risks and the probability of catastrophic events occurring and educating them to be proactive in order to increase independent choices oriented towards prevention.

In the case of existing exposure reduction purposes, the main approaches are of incentive and regulatory-coercive type.

The *incentive approach* is based, for those who have already chosen to live or work in a hazardous place, on the inducement to transfer houses and workspaces to a safer place;

The *regulatory-coercive approach* is based on the use of extraordinary means and powers and consists of declaring certain areas no longer habitable with the consequent delocalisation of the settled population and/or economic activities.

Through the combination of these two approaches it is possible to transfer existing buildings (i.e. building rights already exercised) through equalisation or compensation spatial planning tools. It also possible to remove existing buildings, ensuring the restoration of the natural pre-condition of the site afterwards, and to transfer the settled population and economic activities to safer places.

Both in case of potential and existing exposure, hazard avoidance strategies are "unlikely to be viewed favourably by affected property owners because of the potential loss in investment in the land" (Turbott and Stewart 2006, p. 3). Intuitively, regardless of the approach adopted, these kinds of actions tend to collide with some civil and constitutional principles. To be effective they need first of all to be grounded on a deep awareness and clear perception of risk. This is due to the fact that both strategies are based on the application of different types of rules (juridical, moral, social) and have a strong symbolic value since they are linked to apparently inviolable rights such as property right and right to housing. They both require high public investments in order to increase knowledge, management, planning and control capacity. Moreover, hazard avoidance strategies are negatively affected by social and political predisposition to illegality and corruption.

7.4   The Key Role of Land Use Planning and Buildings Bans

The best long-term mechanism of modern-day hazard mitigation is land use planning (Cronin and Cashman 2007, p. 196).

Safety promotion and risk prevention are normally based on the "ability to predict and identify differential risk areas" and to match them with a regulatory apparatus. Within hazardous areas, regulatory apparatuses should state cautionary and structural strengthening actions and provide clear prohibitions, constraints and land use limitation systems (Di Sopra 2017, p. 80).

Given data reported in available repositories it looks like there is a tendency of disasters to increase all over the world and for this reason they represent one of the most urgent issues that modern societies are called on to face. Although today there is a tendency to attribute this increase to climate change, some authors point out that the first cause of natural disasters is to be found in the rapid growth of the world population and in the consequent increase in human exposure to natural phenomena due to the urbanisation of highly dangerous areas (Grattan and Torrence 2007).

As already explained, when a highly difficult to predict (as for timing and location), uncontrollable and destructive natural event may occur in a limited territory, the unconditional prohibition of settling new population appears as the most effective way to mitigate risk. This is especially true as regards the volcanic scenario because of the precise spatial identification of the hazard source (Pierson et al. 2014). However, this type of operations encounter a series of difficulties, first of all with the need to make hazard measurements and calculations as objective and precise as possible; secondly because, in order to achieve quantitative estimations of the exposed value— that vary in time and space—asset values have to be estimated for all the affected objects (Merz et al. 2010). Furthermore, it should be noted that the more distant the last catastrophic event occurred in the memory of local populations, the more difficult it is to bring public opinion to favour decisions that risk appearing excessively severe and therefore unpopular. Regulatory policies built around "absolute prohibitions" are notoriously clashing with people's identity, emotional or cultural attachment to places, but especially with commercial and real estate interests. Nevertheless, in many cases buildings constraints and bans are vital since they can have different specific aims in name of public interest such as landscape and ecological tutelage, hydrogeological stability and water regime safeguard, human safety and health protection, capital facilities and assets defence. In most cases the restrictions imposed on urbanisation, although specifically aimed at protecting the environment and the landscape (also through the establishment of protected areas), indirectly pursue disasters prevention through the containment of anthropic exposure to natural risks.

7.5 Volcanic Hazard-Avoidance Strategies and the Italian Case

Over the past centuries all around volcanoes human activities have not taken into account the problem of volcanic hazard leaving a legacy of high risk situations today (Rosi 2000). It was only in the second half of the twentieth century that volcanic hazard became a fundamental issue also in terms of risk prevention because of the rapid growth of victims recorded during and after volcanic eruptions. In recent decades, the attention of the scientific community towards volcanic risk has been progressively shifting from an approach totally focused on the study of hazards towards a "holistic" paradigm that integrates geophysical with human and social components. This shift has encouraged scientists and authorities to consider even the psycho-perceptive aspects related to volcanic risk (Leone and Lesales 2009). A contribution to this paradigm change came from the gradual conviction that exposure reduction is the result of various factors starting from correct information and perception of risks by people who live in proximity to highly dangerous volcanoes (*ivi*).

With regard to volcanic scenarios, one of the central issues related to the possibility of implementing exposure containment and reduction policies regards the risk acceptance degree on the basis of hazard maps and scenarios proposed by scientific studies (probability, type and intensity of potential volcanic eruptions[3]). A second issue concerns the spatial precision of areas to be subjected to building restrictions in order to prevent volcanic risk. Rarely can hazard maps actually be so precise as to avoid any ambiguities or uncertainties, but only undeniable hazard perimeters can justify mandatory restrictions on land use without generating potential injustices and litigation. However, if the main task of the scientific community is to define accurate hazard maps, it is the task of politics and public administration to equip local communities with tools—in particular with land use regulation plans—capable of translating hazard and risk maps into clear and consistent rules aimed at limiting human settlement at least in potentially lethal areas.

In Italy, volcanic risk has been managed mainly according to an emergency approach and relying heavily on the technical-scientific ability to capture precursor signals capable of modifying the alert levels and, in extreme cases, of activating emergency and evacuation procedures. However, as the National Civil Protection points out, no form of deterministic forecast is conceivable and it is not possible to predict with absolute certainty when a volcanic eruption will occur or the characteristics it will have. It is also due to these uncertainties that, especially around active volcanoes, in Italy and abroad (Table 7.2) land use limitations have been indirectly entrusted through the institution of natural protected areas and by the implementation of environmental and landscape protection plans. It is no coincidence that the two most well-known Italian volcanoes are also characterised by the presence of two natural parks: the Vesuvius National Park and the Etna Regional Park.[4]

In Italy there are many inhabited areas exposed to volcanic hazards: the circumvesuvian belt, the Piana Campana east of Vesuvius, the Phlegraean Fields and the city of Naples, the islands of Ischia, Stromboli, Lipari and Vulcano, the inhabited belt

---

[3] A volcanic eruption may cause multiple threats such as pyroclastic flows, lava flows, lahars, ash fallout and lapilli fallout (Zuccaro et al. 2008; Thierry et al. 2008).

[4] Among its objectives the Plan of the Vesuvius National Park has the containment of urbanisation, the mitigation of risks connected to seismicity and volcanism, and the opposition of unauthorised construction.

around Etna and the city of Catania. Among the Italian dormant volcanoes –volcanoes "whose current rest time is less than the longest rest period previously recorded"— there are Colli Albani, Campi Flegrei, Ischia, Vesuvio, Lipari, Vulcano, Panarea, Isola Ferdinandea and Pantelleria.[5] Among these, Vesuvius, which has a very low

Table 7.2 International strategies to mitigate volcanic risk through the land use regulations

Case/volcano	Country	Type	Hazard-avoidance strategy
Mount Rainier	USA	Quiescent stratovolcano	Limitation of urban sprawl areas, housing densities and the size and capacity of tourist facilities; prohibition of the location of strategic, sensitive or dangerous structures (hospitals, schools, police or fire-fighters stations, chemical plants, etc.)
Soufriere Hills	Caraibi (UK)	Active stratovolcano	Division of Montserrat Island into three zones: *Exclusion Zone,* where access is prohibited except for scientific monitoring or for national security reasons; *Central Zone,* where there is a single residential area whose residents are in a constant state of alert; *Northern Zone,* where the risk is significantly lower and therefore residence and trade are allowed
Rotorua	New Zealand	Lake in caldera	*Lakes protection belt* where to limit development and housing densities
Taranaki (Mt. Egmont)	New Zealand	Quiescent stratovolcano	*National Park* as a 'tool' to impose building restrictions because of natural land and the ecosystem preservation

(continued)

---

[5]Not all dormant volcanoes present the same level of risk, both for the danger of the expected phenomena and for the different size of the exposed population. See the webpage of the Italian

Table 7.2 (continued)

Case/volcano	Country	Type	Hazard-avoidance strategy
Mount Pelée	Martinique (France)	Active stratovolcano	Preparation of a *Plan de Prévention des Risques* (PPR) which has favoured a strategy of differentiated land use limitations according to the different eruptive scenarios. Risk is declared "acceptable" in areas where eruptive scenarios are less serious and less probable while construction is prohibited only in cases of serious danger and high probability. The PPR proposed a zoning scheme with different land use regulations: non-building areas; building areas according to specific regulations; building areas following some recommended preventive measures; building areas without any prescription
Mount St. Helens	USA	Active stratovolcano	*Volcanic Monumental Park* as an opportunity for: acquiring land within the park boundary (by donation, purchase or exchange with the owners); providing alternative opportunities to residence (tourism, recreational or training activities, study activities and scientific research)
Mt Usu, Hokkaido	Japan	Active stratovolcano	*Perimetration of high risk areas* where human settlement and the location of strategic/sensitive functions are prohibited

(continued)

Table 7.2 (continued)

Case/volcano	Country	Type	Hazard-avoidance strategy
	Indonesia		*Zoning map*: for volcanoes with short quiescent times, three categories of zoning are used: precluded areas; danger areas; alert areas. For volcanoes with long quiescent times, only the last two zoning categories are used

Source Produced by the author and based on different sources

eruptive frequency and is in blocked conduit conditions, is the Italian volcano around which most communities have settled (more than 600,000 people in the most densely populated area of Italy). Vesuvius is considered one of most worrying case in the world because of both its potential eruptive intensity and the enormous exposure in demographic terms within its surrounding areas[6] (cf. Strader et al. 2015).

In 2003 the Campania Region promulgated a specific law (LR 21/2003) that has imposed the so-called "residential block" to the municipalities included in the Vesuvius Red Zone (maximum danger zone) which, for local urban planning schemes and tools, consists of the prohibition of admitting building increase or changes for residential purposes. The same law also prohibits the legalisation of unlawful buildings ("building amnesty block") while in exceptional cases allows the change of housing into production, commercial and tourist activities.

While the task of limiting potential exposure has been entrusted mainly to territorial and landscape-environmental planning, that of reducing the existing exposure through the so called 'housing decompression' strategies has been entrusted to quite sophisticated proposals aimed at relocating resident population through self-determined processes at the local scale (this is the case in particular with Vesuvius). We use here the expression 'housing decompression' (translation of the Italian expression *decompressione abitativa*) referring to government-supported and risk-related managed retreat strategies aimed at relocating households or communities due to widespread hazards.[7] In any case, this type of proposal, developed mainly since the 2000s, was founded on the awareness that the incentives to leave the most

---

National Civil Protection dedicated to the Italian active volcanoes: https://www.protezionecivile. gov.it/jcms/it/vulcani_attivi.wp.

[6] See the webpage of the Italian National Civil Protection dedicated to Vesuvius volcanic risk: https:// www.protezionecivile.gov.it/jcms/it/rischio_vulcanico_vesuvio.wp.

[7] Concepts such as "managed retreat", "planned relocation" and "preventative resettlement" can create some confusion because they are more and more associated with climate change adaptation (Dannenberg et al. 2019) even though they are also suitable for application to different natural and anthropogenic scenarios (cf. Greiving et al. 2018).

hazardous volcanic areas should be supported by residential relocation strategies (see next section). One of the strengths of these 'decompression' strategies has been the financial evaluation of the proposed operations and the proposal of new compatible uses for the evacuated areas. Among the most frequent is the proposal to replace residential areas with urban parks, tourist or recreational structures capable of contributing to limiting the economic damage caused by the loss of population and productive activities. In rare cases, as happened at the turn of the 60s and 80s in Pozzuoli (Campi Flegrei area), mixing the case of the emergency with that of prevention, we saw forced evictions directly conducted by the State followed by the construction of new public housing estates where the population has been permanently relocated. However, these experiences have left an indelible mark on the local populations who still speak of them as 'deportation' acts.

7.6 'Housing Decompression' Strategies for the Vesuvius Red Zone

The *Progetto Vesuvìa—La scelta possibile* (Vesuvìa Project—The possible choice) represents the first attempt to implement a unitary policy of relocation of the Vesuvian populations residing in areas with high volcanic risk. The project was launched in 2002 by the Campania Region to initiate a policy of incentives to reduce the anthropic presence in the Mount Vesuvius hinterland and in particular in the Red Zone (18 municipalities for a total of about 600,000 inhabitants[8]). Through two calls approved by the Regional Council in 2003 and 2004, the Project proposed to encourage a "spontaneous exodus" through economic incentives, consensual measures and urban recovery schemes within the regional territory (Campania Region 2004). Part of the budget was allocated to households who had been renting a house for at least five years in the Red Zone as an incentive to purchase new dwellings outside the Red Zone. Another part of the budget was consequently allocated to support cooperatives and construction companies that were to build new housing or renovated buildings that those households could purchase or lease. A pilot project would have incentivised some families residing in public housing estates to move them to new public housing estates appositely built by the Campania Region with the consequent demolition of the previous ones. Another part of the budget was allocated to encourage property owners to convert their houses into new tourist-accommodation businesses. The smallest part of the budget was allocated to oppose the building of new residences and the conversion of existing ones into productive activities; training activities in schools and for all citizens; the call for the feasibility study of a territorial transformation company, with a public majority, to which to entrust the coordination of the redevelopment of the territory; the functioning of the assembly of mayors of the municipalities involved in the Vesuvius risk mitigation scheme with also the

---

[8]Today the municipalities included in the Red Zone of Vesuvius are 25 for a total of about 700,000 inhabitants.

participation of the Province of Naples and the Vesuvius National Park Authority (cf. Bignami 2010).

After having prepared only 4,000 funding requests, the Vesuvìa Project failed because of financial and political reasons, but also because of real estate market imbalances generated by the unforeseen phenomenon of under-the-table rentals (Meo 2010).

The already mentioned Regional Law n. 21/2003 (so-called "Vesuvius Law") launched the *Piano Strategico Operativo (PSO) per l'Area Vesuviana* (2005–2006).[9] Among other actions, it proposed the activation of schemes for housing density reduction within the Red Zone. The Law provided at the same time a long-term plan and quickly effective actions for the reduction of the exposed population. The plan goal was ambitious: 10% in 15–20 years (about 55,000 inhabitants). In line with the previous Vesuvìa Project, the PSO therefore proposed the conversion of residential buildings into tourist accommodation and residential areas into productive areas. This choice was due to three considerations. First, non-residential buildings would not generate displaced persons looking for a house, with positive consequences on the emergency management and evacuation activities. The PSO was entrusted the task of finding complementary and compensatory solutions to the aforementioned "residential block" in order to mitigate the severity of urban planning legislation by offering citizens new opportunities for development and work (the residential block should not have blocked economic development also). Second, the uncertainty regarding the type of eruptive phenomenon would make any precautionary evacuation of non-residential buildings much more acceptable (a posteriori) than home evacuations. In fact, although the Vesuvian Observatory can record clear warning signals of an eruptive event, it is impossible to accurately predict the nature and extent of events in a period of time sufficient to avoid evacuation or to declare it indispensable. While definite forecasts can be made only a few hours before the eruption, the alarm phase according to Civil Protection must be sounded a few weeks before the eruption in order to evacuate the entire Red Zone. Thus, the proposal to convert residential buildings into buildings with other uses was not dictated by the sole need to minimise potential victims, but also by the need to facilitate evacuations, to avoid humanitarian emergencies and to avoid mass evacuations in the face of not very dangerous volcanic phenomena (possible cry-wolf effect).

The technicians and academics that drew up the PSO were aware of the absence of a real prevention plan for the "Vesuvius risk". Therefore, they had tried, dialoguing with the National Civil Protection, to introduce the concept of *acceptable risk*, thus focusing on the need to "graduate" risk scenarios according to different urban contexts and different types of volcanic phenomena. This perspective would make it possible to open up to more reduction-oriented solutions, not only in terms of exposure but also of vulnerability. This model represented an interesting alternative to models based on the concept of *maximum risk* which contemplate the drafting

[9]For the information provided, we would like to thank Prof. Francesco Domenico Moccia from the University of Naples Federico who was councillor of the Urban Planning Department of the Province of Naples at the time of drafting the PSO.

of evacuation plans as the only possible solution. Among the virtues of the PSO there is also the multi-hazard approach due to a hydrographic component capable of amplifying the tragic effects originating from the eruptive phenomena, e.g. *lahars* (mud avalanches composed of water and hot pyroclastic material) which could arise, as already happened in 1631, from the impact of the volcanic event on the existing hydrographic system. Another peculiarity of the PSO is the assumption that incentives can work only in the presence of new and coordinated regulations which should be as much as possible the result of proposals coming "from below" and locally connoted.

The PSO, approved by the Province of Naples (today Città Metropolitana di Napoli) in conformity with its Provincial Spatial Planning Scheme (PTCP), was never approved by the Campania Region probably because it was considered too binding with respect to the use that the Region should have made of the European Structural Funds.

7.7   Learning from Vesuvius: When and Why 'housing Decompression' Strategies Do Not Work

Risk and uncertainty are central to the process of institutional defence, with cultural biases orienting people's selection of dangers to accept or to avoid, judgements about fairness of distribution of risks across society, and who to blame when things go wrong

(Bammer and Smithson 2008, p. 352).

If prevention policies based on land use regulation seem to be effective in relation to potential exposure, 'housing decompression' strategies appear extremely complex, not very incisive in quantitative terms and often unfeasible even in the face of extreme concrete risks. Looking at the case of the Vesuvius, it is not easy to evaluate the actual effectiveness of the 'housing decompression' plans because none of them has ever reached a sufficient degree of implementation. Despite this, the ways and reasons for which the implementation has found difficulties equally allow us to express some considerations on the biased fallacy of this type of initiative. One of the main problems of 'housing decompression' strategies like the Vesuvìa Project and Vesuvian PSO projects is represented by a vision totally focused on the transfer and conversion of residential properties without contemplating the social and cultural difficulties in relocating people and business activities outside their original fabric and milieu. Another limit, revealed in particular by the very partial implementation of the Vesuvìa Project, was the moral hazard behaviour of incentivised households who illicitly rented their former homes—that were supposed to be transformed or removed—to other households, thus creating new exposure. Moral factors are strongly related to perceived risk (Sjöberg and Torell 1993) and risk perception plays a crucial role in hazard avoidance strategies. Hazards in general "are conceived as such, rather than as the outcome of human actions" (Sjöberg 2000, p. 4), but volcanic hazard is one of the most natural (thus less immoral) and far from *hazardscape* interpretation.[10] This is particularly important if we consider risk perception as a function of properties of the hazards (*ivi*, p. 9). The fact that immoral risks are perceived as more severe helps us also to understand why individual behaviours and activities that seem not to damage other people directly are perceived as less worrying. But this is not sufficient to explain the failure of housing decompression strategies. Daily tangible hazards and those that have left an imprint in the population are the most visible, while spatially selective or infrequent hazards are less visible to exposed people (Ley-García et al. 2015). The nature of the *putative harm* (Hilgartner 1992) enables us to explain why volcanic risk, while being inherently better treatable through hazard-avoidance strategies, shows such a degree of fallacy from the point of view of long-term prevention policies and in particular of housing decompression policies. In the case of Vesuvius, eruptions are "known events [...] judged to have a negligible probability of occurrence, and thus not believed to occur" (Motet and Bieder 2017, p. 28, referring to Aven and Krohn 2014). It is as if, for certain remote risks, even when they are well spatialised, prevention policies struggle to work due to the lack of "anticipatory frames" (Vogel 2008) and true "representations of the future" (Brown and Michael 2003), while the prefiguration of critical scenarios becomes a quite de-politicised achievement functional only to manage the emergency rather than to mitigate the risk. In the case of volcanic risk exposure emergency management strategies are in fact usually preferred since they can be implemented effortlessly through alert systems and evacuation plans rather than preventing and reducing exposure structurally.

Risks are socially defined (Beck 1992), inherently ambiguous and subject to "constrained relativism" (Thompson et al. 1990; see also Tansey and O'Riordan 1999). Thus, even where real prevention strategies could be applied suitably, they tend to not be implemented for reasons often linked to political consensus but also to socio-cultural biases. Cultural theorists of risk have stated that behaviours and attitudes towards risk and hazard vary systematically within different contexts and societies essentially according to four cultural biases (or world views): *individualism, fatalism, hierarchism* and *egalitarianism* (Douglas and Wildavsky 1982; Wildavsky and Dake 1990). In the case of the people living around Vesuvius, fatalism and individualism seem to prevail with respect to risk perception (see also Rippl 2002).

A last issue to raise here regards the indirect perceptive effects of mitigation plans and projects. The case of Vesuvius shows that decompression projects themselves can be perceived as potential threats by the involved communities. This is due to the fact that also "regulatory approaches to risk can themselves create risks of their own" (Francis and Knott 2014, p. 92). This is something not so far from the concept of "risk perception shadow" coined to describe the locally affected population that perceives itself to be at risk from a proposed project (Stoffle et al. 1993). Therefore, if we assume that the "perception of hazardscape is heterogeneous and does not match

---

[10]The expression "hazardscape" was originally coined in the field of technological hazards (Corson 1999). In an ecological perspective of risks it is intended as "the place and people's characteristics that favour hazard occurrence in particular spatio-temporal contexts" (Khan 2012, p. 3775).

with objective exposure" and that "biased perceptions of hazards (due to amplification or invisibility) [...] may lead to erroneous judgments concerning safety" (Ley-García et al. 2015, p. 494), we can conclude that hazard-avoidance strategies—and 'housing decompression' policies in particular—should work socially and culturally on risk perception in order to align it with actual hazard and exposure, stimulate more conscious and independent choices of people and facilitate the implementation of long-term plans and projects. Even when abundant and detailed engineering and administrative information is available, people "tend to close themselves in their universe of meanings and symbols" (Colombo 1995, p. 93, own transl.). Since the understanding of the *community character* play a crucial role in volcanic risk communication strategies (Andreastuti et al. 2019) the sources of information should be first of all familiar and visible with respect to the specific cultural context of the social target (Colombo 1995).

7.8 Conclusion

Albeit in a non-exhaustive way, this contribution has shown the extreme complexity of public policies aimed at reducing residential exposure to natural risks. These kinds of policies seem to be inherently ineffectual especially when meant to reduce existing housing exposure in situations of volcanic hazards. With specific reference to the Vesuvian riskscape, the experimental nature of the solutions adopted—when not merely emergency-oriented—has produced interesting projects but too weak with respect to risk perception issues and to inevitable political, social and cultural resistance and biases. These obstacles have led also to the failure of potentially virtuous 'housing decompression' plans and schemes. A final suggestion is to improve the effectiveness of these projects by working more rigorously on local community perception, expectations and involvement.

References

Andreastuti S, Paripurno E, Gunawan H, Budianto A, Syahbana D, Pallister J (2019) Character of community response to volcanic crises at Sinabung and Kelud volcanoes. J Volcanol Geotherm Res 382:298–310. https://doi.org/10.1016/j.jvolgeores.2017.01.022
Aven T, Krohn BS (2014) A new perspective on how to understand, assess and manage risk and the unforeseen. Reliab Eng Syst Saf 121(1):10. https://doi.org/10.1016/j.ress.2013.07.005
Bammer G, Smithson M (eds) (2008) Uncertainty and risk: multidisciplinary perspectives. Earthscan, London
Barca F, Casavola P, Lucatelli, S (2014) Strategia nazionale per le aree interne: definizioni, obiettivi e strumenti di governance. Materiali Uval 31
Beck U (1992) Risk society: towards a new modernity. Sage, London
Bignami D (2010) Protezione civile e riduzione del rischio disastri: metodi e strumenti di governo della sicurezza territoriale e ambientale. Maggioli, Rimini

Brown N, Michael M (2003) A sociology of expectations: retrospecting prospects and prospecting retrospects. Technol Anal Strat Manag 15(1):3–18. https://doi.org/10.1080/095373203200004 6024

Burby RJ (1998) Cooperating with nature: confronting natural hazards with land-use planning for sustainable communities. Joseph Henry Press, London

Burby RJ, Dalton L (1994) Plans can matter! The role of land use plans and state planning mandates in limiting the development of hazardous areas. Publ Administr Rev 54(3):229–238. https://doi.org/10.2307/976725

Casa Italia—Struttura di Missione (2017) Rapporto sulla promozione della sicurezza del patrimonio abitativo. Italian Presidency of The Council of Ministers, Casa Italia Department

Colombo M (1995) Convivere con i rischi ambientali: il caso Acna, valle Bormida. FrancoAngeli, Milano

Corson M (1999) Hazardscapes in reunified Germany. Glob Environ Change Part B: Environ Hazards 1(2):57–68. http://dx.doi.org/10.1016/s1464-2867(99)00009-1

Cronin SJ, Cashman KC (2007) Volcanic oral traditions in hazard assessment and mitigation. In: Grattan J, Torrence R (eds) Living under the shadow: cultural impact of volcanic eruptions. Left Coast Press, Walnut Creek, CA, pp 175–202

Dannenberg AL, Frumkin H, Hess JJ, Ebi KL (2019) Managed retreat as a strategy for climate change adaptation in small communities: public health implications. Clim Change 153(1–2):1–14. https://doi.org/10.1007/s10584-019-02382-0

De Rossi A (2018) (ed) Riabitare l'Italia: le aree interne tra abbandoni e riconquiste. Donzelli, Rome

Di Sopra L (2017) Confronto dei modelli di ricostruzione: verso una legge quadro nazionale. In: Fabbro S (ed) Il 'Modello Friuli' di ricostruzione. Forum Udine, pp 61–83

Douglas M, Wildavsky A, (1982) Risk and culture: an essay on selection of technological and environmental dangers. California University Press, Berkeley

Esteban JF, Izquierdo B, Lopez J, Molinari D, Menoni S, Roo A, De Eftichidis G (2011) Current mitigation practices in the EU. In: Menoni S, Margottini C (eds) Inside risk: a strategy for sustainable risk mitigation. Springer, Milan, pp 129–186

FEMA—Federal Emergency Management Agency (1995) FEMA—Federal Emergency Management Agency 1995 National mitigation strategy FEMA, Washington DC

Francis M, Knott K (2014) Aum Shinrikyo and the move to violence. In: UK Government, Chief Scientific Adviser (ed) Innovation: managing risk, not avoiding it: evidence and case studies. UK Government, November 2014

Grattan J, Torrence R (eds) (2007) Living under the shadow: cultural impact of volcanic eruptions. Left Coast Press, Walnut Creek, CA

Greiving S, Du J, Puntub W (2018) Managed retreat: a strategy for the mitigation of disaster risks with international and comparative perspectives. J Extr Events 05(02–03):1850011. https://doi.org/10.1142/s2345737618500112

Hilgartner S (1992) The social construction of risk objects: or, how to pry open networks of risk. In: Short Jr. JF, Clarke L (eds) Organizations, uncertainties and risk. Westview Press, Oxford

Khan S (2012) Disasters: contributions of hazard scape and gaps in response practices. Nat Hazards Earth Syst Sci 12:3775–3787. https://doi.org/10.5194/nhess-12-3775-2012

Leone F, Lesales T (2009) The interest of cartography for a better perception and management of volcanic risk: from scientific to social representations: the case of Mt. Pelée volcano, Martinique. J Volcanol Geotherm Res 186(3–4):186–194. https://doi.org/10.1016/j.jvolgeores.2008.12.020

Ley-García J, Denegri FM, Ortega LM (2015) Spatial dimension of urban hazardscape perception: the case of Mexicali. Mexico Int J Disast Risk Reduct 14(4):487–495. https://doi.org/10.1016/j.ijdrr.2015.09.012

Menoni S (2019) Per un nuovo approccio alle strategie e agli interventi di prevenzione e riduzione dei rischi naturali: applicazione al caso della ricostruzione post-terremoto. In: AA. VV. Atti della XXI Conferenza Nazionale SIU: confini, movimenti, luoghi: politiche e progetti per città e territori in transizione, Firenze, 6–8 giugno 2018. Planum Publisher, Roma, Milano, pp 1276–1282

Meo S (2010) Vesuvio sicuro, fallito il piano di Bassolino. Il Sole 24 Ore (27.07.2010). Available online https://st.ilsole24ore.com/art/notizie/2010-07-27/fallimento-piano-vesuvia-160157_PRN.shtml. Last access 11.03.2020

Merz B, Kreibich H, Schwarze R, Thieken A (2010) Review article "Assessment of economic flood damage". Nat Hazards Earth Sys Sci 10(8):1697–1724. https://doi.org/10.5194/nhess-10-1697-2010

Motet G, Bieder C (2017) The illusion of risk control. Springer, France

NOAA Coastal Service Center (1999) Community vulnerability assessment tool. New Hanover County, North Carolina. Available online: https://geodata.lib.ncsu.edu/fedgov/noaa/commvuln/startup.htm

Pesaro G, Mendoza MT, Minucci G, Menoni S (2018) Cost-benefit analysis for non-structural flood risk mitigation measures: insights and lessons learnt from a real case study. Saf Reliab Safe Soc Chang World 109–118. https://doi.org/10.1201/9781351174664-14

Pierson TC, Wood NJ, Driedger C (2014) Reducing risk from lahar hazards: concepts, case studies, and roles for scientists. J Appl Volcanol 3(16). https://doi.org/10.1186/s13617-014-0016-4

Regione Campania—Presidenza della Giunta e Assessorato all'Urbanistica (2004) La scelta possibile: guida alle opportunità del progetto regionale VESUVIA per i cittadini della zona a più alto rischio vulcanico. Regione Campania, Naples

Rippl S (2002) Cultural theory and risk perception: a proposal for a better measurement. J Risk Res 5:147–165. https://doi.org/10.1080/13669870110042598

Rosi M (2000) Il rischio vulcanico in Italia. Seminario nazionale per la diffusione della cultura della protezione civile nella scuola dell'obbligo. Ministero dell'Interno e Ministero della Pubblica Istruzione, Rome. Available online http://www.icvoltalatina.gov.it/Sicurezza/04%20Il%20risc hio%20vulcanico.pdf. Last access: 11.06.2017

Sjöberg L, Torell G (1993) The development of risk acceptance and moral valuation. Scand J Psychol 34:223–236. https://doi.org/10.1111/j.1467-9450.1993.tb01117.x

Sjöberg L (2000) Factors in risk perception. Risk Anal 20(1):1–12. https://doi.org/10.1111/0272-4332.00001

Stoffle RW, Stone JV, Heeringa SG (1993) Mapping risk perception shadows: defining the locally affected population for a low-level radioactive waste facility in Michigan. Environ Profess 15(3):316–333. https://doi.org/10.1525/aa.1991.93.3.02a00050

Strader SM, Ashley W, Walker J (2015) Changes in volcanic hazard exposure in the Northwest USA from 1940 to 2100. Nat Hazards 77:1365–1392. https://doi.org/10.1007/s11069-015-1658-1

Tansey J, O'Riordan T (1999) Cultural theory and risk: a review. Health Risk Soc 1(1):71–90. https://doi.org/10.1080/13698579908407008

Thierry P, Stieltjes L, Kouokam E, Nguéya P, Salley PM (2008) Multi-hazard risk mapping and assessment on an active volcano: the GRINP project at Mount Cameroon Nat Hazards 45:429–456. https://doi.org/10.1007/s11069-007-9177-3

Thompson M, Ellis R, Wildavsky A (1990) Cultural theory. Westview Press, Boulder CO

Turbott C, Stewart A (2006) Managed retreat from coastal hazards: options for implementation. Environment Waikato Technical Report 2006/048

UNDHA (1992) Internationally agreed glossary of basic terms related to disaster management. United Nations Department of Humanitarian Affairs. Geneva/New York

UNISDR (2009) UNISDR Terminology on disaster risk reduction. United Nations International Strategy for Disaster Reduction (UNISDR), Geneva. Available online at https://www.preventio nweb.net/files/7817_UNISDRTerminologyEnglish.pdf

UNDRR (2019) Global assessment report on disaster risk reduction. United Nations Office for Disaster Risk Reduction (UNDRR), Geneva

Varnes DJ (1984) Landslide hazard zonation: a review of principles and practice United Nations Educational Scientific and Cultural Organisation, Paris

Vogel KM (2008) 'Iraqi Winnebagos of death': imagined and realised futures of US bioweapons threat assessments. Sci Publ Pol 35(8):561–573. https://doi.org/10.3152/030234208x377407

Wildavsky A, Dake K (1990) Theories of risk perception: who fears what and why? Daedalus 4:41–60
Zuccaro G, Cacace F, Spence R, Baxter P (2008) Impact of explosive eruption scenarios at vesuvius. J Volcanol Geoth Res 178:416–453. https://doi.org/10.1016/j.jvolgeores.2008.01.005

Chapter 8
Urban Resilience as New Ways of Governing: The Implementation of the 100 Resilient Cities Initiative in Rome and Milan

Alessandro Coppola, Silvia Crivello, and Wolfgang Haupt

8.1 Introduction: Resilience Thinking and the New Policy Orthodoxy

The discourse on urban resilience has become extremely influential around the globe, as shown by extensive contributions to the academic debate, including urban studies (see, for example, Vale and Campanella 2005; Meerow et al. 2016; Coaffee 2016; and Mehmood 2016). In the context of an unstable and turbulent world, cities are seen both as bundles of extremely valuable assets, particularly exposed to all sorts of risks, and as sources of unfettered skills and capacities that must be proactively mobilised to face those risks (Jabaree 2013). Going beyond the entrenched debates on emergency management, supporters of urban resilience have attempted to reframe how cities face risks by reconceptualising and linking discussions on preparedness and recovery. In this perspective, the ability of local systems to be effectively prepared for the manifestations of certain risks, to mitigate their impacts and to improve their ability to recover from them have been seen as essential dimension of urban resilience. This discussion has also, however, spurred heated debates. Initial discussions featured differences between the 'bouncing back' and 'bouncing forward' approaches to resilience (see, for example, Shaw and Theobald 2011); the former is more in line with previous engineering approaches, and the latter is more in line with

A. Coppola (✉)
Department of Architecture and Urban Studies, Politecnico di Milano, Milan, Italy
e-mail: alessandro.coppola@polimi.it

S. Crivello
Interuniversity Department of Regional and Urban Studies and Planning, Politecnico di Torino, Turin, Italy
e-mail: silvia.crivello@polito.it

W. Haupt
Leibniz Institute for Research on Society and Space, Erkner (IRS), Erkner, Germany
e-mail: wolfgang.haupt@leibniz-irs.de

socio-ecological approaches and the vision of a plurality of possible outcomes and time frames (Baravikova et al. 2020). Notwithstanding these different approaches, critical readings have dubbed urban resilience a 'fuzzy concept' whose consistency with neo-liberal thinking (Davoudi and Porter 2012) has been stigmatised (Coaffee 2013a, b; Welsh 2014) as it strives to place more responsibility on individuals and collectivities in managing evolving risk configurations that largely depend on societal factors. At the same time, the concept has also been connected with wider processes regarding the 'neoliberalisation of nature' (Heynen and Robbins 2005; Pellizzoni 2015) associated with the spread of new calculative habits and devices aimed at managing growing uncertainty through decentralising and responsibility assigning strategies. Often vaguely defined and uncertainly implemented, under the label of urban resilience policies and strategies cities have advanced a multitude of interventions (see McEvoy et al. 2013; Sharifi et al. 2017; Baravikova et al. 2020). Among others, these include policies as diverse as strengthening the ability to manage migration fluxes (Coaffee 2013a, b), removing pavement from streets to make urban soil more permeable and adaptive to intense precipitation and flooding brought about by climate change (Restemeyer et al. 2015), strengthening economic diversity in the face of changes in global markets and supply chains (Viitanen and Kingston 2014), and enhancing the management of critical assets through data-intensive and smart-city strategies (Rosato et al. 2018). But besides advancing a variety of specific interventions, the discourses on urban resilience featured in policy communities and networks of practitioners have also advanced an understanding of it as a set of qualities and capacities inherent to cities as systems that can be powerfully enhanced through specific actions.

A new urban policy orthodoxy seems to have been established. With clear linkages to previous discourses on strategic planning, this orthodoxy has encompassed elements that mostly revolve around issues of governance. In this regard, cities must become the locus of new strategic planning operations able to encompass different issues and bridge different policy communities (see, for example, Mell et al. 2017). As emphasised in debates on urban policy mobilities, cities need to establish evidence-based policies, especially through alliances with all sorts of knowledge providers (see, for example, McCann 2011a). It is argued that cities need to take a step away from established ways of governing by generalising the use of 'governance networks' as the standard way to make policy (Owens et al. 2006).

As with other paradigms such as urban sustainability, the emerging discourse on urban resilience has been vastly mobilised in international networks through mechanisms of imitation, policy circulation, policy transfer, and policy mobility (Fisher 2014; Haupt et al. 2019; Croese et al. 2020). Some international organisations, in particular transnational municipal networks (TMNs), have taken the lead in this process and—by providing detailed guidelines and promoting supposedly best practices—have strived to transfer broad understandings of urban resilience to different local contexts (Hakelberg 2014; Busch 2015; Haupt and Coppola 2019). Scholars working on urban resilience, urban governance, and urban policies have scrutinised these processes (Evans 2011; Davoudi and Porter 2012). In particular, it is critically important to study how urban resilience discourses circulating in international networks are concretely translated at the local level, how certain local factors contribute to shaping outcomes, and how network-organisations, and the policy mobilities they promote and facilitate adjust to local differences.

Based on this broad agenda, this paper focuses on the mainstreaming and institutionalisation of discourses on urban resilience within local governments and governance networks through the intervention of TMNs. Specifically, the study analyses the implementation of the US-based international initiative 100 Resilient Cities (100RC) in the only two Italian member cities, Milan and Rome. The cases appear particularly relevant as the initiative is relatively new (the programme started in 2014), and—as it promotes the establishment of new civil servant figures leading an urban resilience office—it implies a strong focus on changing local government and governance structures. In this sense, we can look at institutionalisation and mainstreaming through policy mobility in the early days as we study processes that are institutionalising urban resilience as a legitimate area of policy intervention in the context of developing a resilience strategy (meaning that policy institutions and actions are institutionalised contextually). Moreover, the focus on Italian cases allows us to inquire into these processes in two different cities in which factors such as political institutions and multi-level governance systems are similar.

The paper does not question the quality of the initiative, its implementation, nor its effects in terms of urban resilience. Rather, it starts from an essential, overarching question: how is the 'one size fits all' and strongly governance-oriented approach of 100RC to urban resilience actually implemented on the ground, and with which outcomes?

This paper is organised as follows. First, we briefly introduce and discuss the literature on urban policy mobilities, TMNs, and network governance. Then we present the essential characteristics of the 100RC initiative and of its rationales. Then we present and discuss the evidence collected for the two cities, introducing three main dimensions of analysis, and provide a description of the methodology and materials employed. Finally, we offer some conclusive thoughts.

8.2 Policy Mobilities, Transnational Municipal Networks, and Local Governance Networks

The expanding role of international activities in city governments has been linked to state rescaling processes, changes in capitalist regulation, and the spread of entrepreneurial urban development models (see, for example, the classic papers by Harvey 1989; Cox 1993; Hall and Hubbard 1996; Brenner 2004). Resources appropriated in the context of such activities have become particularly important at a time when mayors' legitimisation strategies have changed from a 'politics logic' based on rewarding localised constituencies through targeted and more traditional interventions to a 'policy logic' based on wider strategies aimed at addressing urban problems by appealing to the general public (Béal and Pinson 2014). This shift has been linked to the need for new ways to govern highly fragmented urban societies by involving a wider range of specialised stakeholders and professional communities in policy-making processes. Scholars researching what is called 'network governance' have described it as systems of 'mutual interactions between a variety of interdependent actors, each with their own motives, who come together to solve a common problem' (Khan 2013, pp. 137–138). Studies in this area have focussed on these changes, acknowledging the growing importance of diverse actors in contemporary urban governance, challenging the traditional understanding of government with the state as the main, if not only, regulator. In this vein, more attention has been paid to forms of looser governance with a more diverse set of actors, such as NGOs or civil society, private enterprises, and knowledge institutions (Pierre and Peters 2000; Bogason and Musso 2006; Sørensen 2014). Critical readings of these developments have under-scored how this variety of actors includes only a small number of democratically legitimised officials, raising concerns about a form of urban policy-making that is increasingly influenced by non-elected elites (Bulkeley et al. 2003; Khan 2013); how the promised flattening of power imbalances and the decline of the central role of the state predicted by so-called post-traditionalist readings of networked governance have not truly materialised (Davies 2011); and how such models both depend on and imply a high level of mobilisation of the middle classes with the likely reproduction of pre-existing power imbalances in urban policy arenas (Caudo and Coppola 2020).

In this context, scholars of 'policy mobilities' have focussed on the different ways urban policies are formed in a closely interconnected, increasingly mobile, and trans-local world (see, for example, Marsh and Sharman 2009; Peck and Theodore 2010; Prince 2010; McCann 2011b; Crivello 2015). The literature in the field analyses how urban policies move from place to place, how they transform spaces, and how policies change during their travels (see Stone 2004; Peck and Theodore 2010; Tenemos et al. 2019). In other words, the literature on urban policy mobility analyses and describes the multiple, complex, and hybrid mechanisms that enable the imitation, mobilisation, and implementation—as well as mutation and adaptation—of policies between a supposedly successful place and another place that imitates it. In academic debates, the concept of policy mobility increasingly replaces the more state-centred and rigid one of 'policy transfer', emphasising how policies mutate during their travels (see McCann 2011).

The circulation of urban policies and the political investment of mayors in inter-national exchanges is more often being actively supported by international organ-isations that act as brokers and 'promote certain approaches within international policy circuits' (Robinson 2015, p. 833). In this context, several scholars have noted the importance of TMNs as providers of platforms for sharing and disseminating knowledge and policies among the member cities (see Kern and Bulkeley 2009; Feldman 2012; Fisher 2014; James and Verrest 2015; Fenton and Busch 2016; Mejía-Dugand et al. 2016; Haupt et al. 2019; Nagorny-Koring 2019). In general, to be defined as such, TMNs must be formal organisations (Kern and Bulkeley 2009; Busch 2015; Haupt and Coppola 2019); they have to create an international secre-tariat and national/sectoral coordinators to run the internal governance; they need a presidency, board, and general assembly responsible for general decision-making; and these must be based on the member cities (Kern and Bulkeley 2009).

The TMNs with a distinct focus on climate governance are also known as 'transnational municipal climate networks' (e.g. Hakelberg 2014; Busch 2015; Haupt and Coppola 2019). Among the TMNs that address local climate action—as 100RC does—the thematic focus is quite heterogeneous. Some TMNs place strong emphasis on ideas of sustainable development (e.g. Local Governments for Sustainability, ICLEI, Citynet, and Cities Development Initiative for Asia), while others focus on climate change mitigation (e.g. C40, Climate Alliance, and Energy Cities), and yet others focus on climate change mitigation and adaptation (e.g. Covenant of Mayors and Compact of Mayors) (Haupt 2018; Haupt and Coppola 2019). Among TMNs that focus on resilience, there are also major differences in their thematic realms. Then again, 100RC views resilience as all-embracing as possible, with climate action being one among many policy areas. Over time, TMNs have been broadening the range of actors involved in their own operations and the kinds of policy-making and governance processes they recommend for their member cities. From their initial focus on local, regional, and national policy-makers, government agencies, and NGOs (Feldman 2012), they have increasingly involved the private sector in the form of companies, consultants, and knowledge providers (Lidskog and Elander 2010; Granberg et al. 2015; Mejía-Dugand et al. 2016; Haupt and Coppola 2019).

In sum, internationalisation has become a key strategic dimension in the restructuring of urban governance because many mayors are eager to appropriate the resources that such activities seem able to supply. These resources have become particularly vital for the legitimisation strategies of mayors and urban governments focussing on policy changes and innovative measures for dealing with emerging urban issues and problems (Pinson and Bear 2014). TMNs have proved to be well positioned to symbolically represent an emerging global city and mayoral leadership on such issues (Barber 2013) and to meet this emerging need by supplying certain resources (Haupt 2018; Haupt and Coppola 2019). While mayoral support and participation have become essential to the growth of such organisations, the support of these organisations has also become critical for mayors, especially mayors who are willing to embark on these leadership strategies by engaging in policy experiments through networked governance exercises. In this context, urban resilience, with its connection to rising urban issues such as climate change and its more general 'urban reform' flair, has become a relevant topic for TMNs. And thanks to the strongly governance-oriented nature of the urban resilience discourse, changes in designing institutional and policy processes have become important enough to become a prime focus of the policy mobility processes embedded in such exchanges. In particular, 100RC has ostensibly focussed on local institutional change and network governance models, as Chief Resilience Officers (CRO)—the new position that 100RC aimed at establishing within city governments—had 'to demonstrate how multiple stakeholder groups—city leadership, civil society, the private sector to name a few—will be actively engaged in building urban resilience' (100RC undated).

We argue that studying the local implementation of such initiatives is a very relevant opportunity to analyse trans-local urban policies *through* institutional change and networked governance in the context of various mayors' legitimisation strategies.

8.2.1 A Network-Like Centralised Organization: Introducing 100 Resilient Cities

Characterised by its long-standing involvement in urban affairs, the New York City-based Rockefeller Foundation (RF) decisively strengthened its focus on issues of climate and urban resilience under Judith Rodin's presidency. Previously an academic dean, Rodin was appointed in 2012 to co-chair a commission aimed at improving New York state's infrastructure resilience and later published *The Resilience Dividend* (Rodin 2014), a book of stories and policy advices related to the resilience of cities. In the meantime, the RF embarked on supporting climate resilience projects in South-East Asia as early as 2008, and in 2013 (the year of its centennial) launched a hundred-million-dollar 'urban resilience challenge' with the promotion of the 100RC initiative. The mission of 100RC was 'helping cities around the world build resilience to the economic, social, and physical challenges that are increasingly part of the twenty-first century' (100RC). Member cities were supported by the provision of finances for creating the position of CRO and were asked to develop resilience strategies. In mid-2019, the RF announced that it had stopped funding the program but had provided $8 million (US) to continue supporting ongoing resilience initiatives in the member cities.

In 2013, 100RC convened a panel of 'leading global figures' to select the first 32 member cities, followed by another 35 in 2014 and 33 more in 2016. Unlike most TMNs, which can be joined by any city that meets the specified requirements (e.g. being in a certain world region, paying the membership fee), to join 100RC a city had to apply for membership and be selected. Based on the classification proposed by Haupt and Coppola (2019), this made 100RC an 'inclusive elite network', as opposed to the 'inclusive mass networks' that constitute the majority (Haupt and Coppola 2019). One more significant difference with other TMNs was the size of the central organisation, as the initiative was rather small in terms of member cities but quite large in terms of the number of employees working for the organisation (Haupt and Coppola 2019). Over time, and not unlike other TMNs, 100RC decentralised its structure and established three regional offices in London, Singapore, and Mexico City in addition to its headquarters in New York, where the vast majority of staff was located. After becoming a member of 100RC, each city was requested to appoint a CRO, which Michael Berkowitz, 100RC's former president, characterised as 'a high-level adviser to the city's chief executive, who has the responsibility of working across sectors of society and silos of government so that key decision-makers connect important strands of work' (Berkowitz 2015).

A key task of the CRO and their team was the development of a city resilience strategy. Cities had first to identify acute shocks (single event disasters) and chronic stresses (regularly recurring disturbing factors) their cities were confronted with (see Fig. 8.1). These shocks and stresses could be very diverse and involve physical, economic and ecological aspects, or a combination of them (see Fig. 8.1 for the distribution of shocks and stresses in 100RC's sponsored urban resilience strategies) and could have different impacts on different components of city systems. In this regard, this approach was consistent with the 'fuzzy', multi-faceted, and difficult to operationalise 'holistic' understanding of urban resilience that is used in many disciplines (see McEvoy et al. 2013; Sharifi et al. 2017; Baravikova et al. 2020). Later, cities were asked to identify thematic areas to be further developed into precise actions aimed at improving the overall resilience of the city as part of a final strategy to be implemented in the following years.

The process would entail a mixture of desk research and 'stakeholder engagement' activities and would be structured around milestones. Many research activities were based on the use of specific tools such as the 'resilience framework' defined by the multinational design and engineering firm ARUP (see Baravikova et al. 2020)—the closest partner of 100RC—and other tools for analysing the vulnerability of critical assets and infrastructures designed by several consultancies. Furthermore, cities would be given the direct support of the organisation with the identification of a 'city relationship manager' from the 100RC staff as well as a 'strategy partner' contracted by the organisation. The 'strategy partner' would usually be a consulting firm or a large NGO—organisations such as the same ARUP, Aecom, ICLEI, or the Rand Corporation—and would support cities in the day-to-day decision-making process of designing the city resilience strategy based on highly defined, standardised procedures. Member cities—and more specifically, CROs—would receive regular visits from 100RC staff and strategy partners and would also meet regularly at a

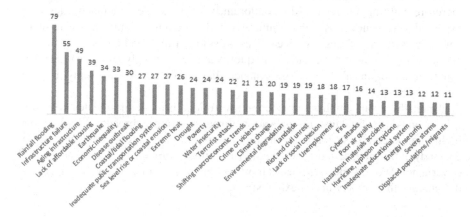

Fig. 8.1 Most common shocks and stresses as identified by the 100RC member cities (showing the number of times the shock or stress was identified by a city). *Source* Our diagram based on 100RC, undated

yearly 'resilience world summit', at regional events, and at thematic network events bringing together cities with comparable needs and priorities. Higher level global mayoral events would take place as well, such as the Mayors Resilience Summit that was held in 2015.

One of 100RC's strategic goals was to set up a marketplace for innovative services for urban resilience. The organisation connected its members to a platform of private partners including large corporations, consultancies, NGOs, and research institutions. Such entities were supposed to support cities in identifying innovative ways to pursue their goals while offering limited services to member cities on a pro-bono basis.

Unsurprisingly, 100RC rapidly became the subject of research. In the past few years, several individual in-depth case studies examined the experiences of 100RC member cities (Sharifi et al. 2017), including Cape Town (Croese et al. 2020), Durban (Sutherland et al. 2019), Rotterdam (Spaans and Waterhout 2016; Dircke and Moolenar 2015), Thessaloniki (Komninos et al. 2018; Pitidis et al. 2018), Byblos (Makhoul 2018), Jakarta (Leitner et al. 2018), Santiago de Chile (Svitková 2018), and Melbourne (Fastenrath et al. 2019; Melkunaite and Guay 2016). In this literature some aspects of the initiative have been scrutinised. Specifically, the strong top-down approach (Fastenrath et al. 2019) and the way the organisation and the tools offered by their private partners appear to exert great influence and thus 'preempt alternative understandings of urban resilience' (Leitner et al. 2018, p. 1276) while 'hollowing out public sector tasks and democratic participation' (Fastenrath et al. 2019, p. 2).

Conversely, less attention has been directed towards 100RC's orientation to changing city governments' institutional settings and organisational styles based on the mentioned push towards working collaboratively 'across policy silos' or distinct areas of government. These ideas were strongly embedded in the strategy design procedures and the tools provided by the network, the Urban Resilience Framework, where 'leadership and strategy', 'integrated planning', and 'empowerment of a broad range of stakeholders' were considered key dimensions for assessing, and therefore building, resilience. Most importantly, as mentioned, member cities were expected to appoint CROs—and ensure they had close ties with mayors—and set up new trans-sectoral offices. The idea of establishing a top-level advisor next to the city mayor—see Berkowitz's statement quoted above—strongly relied on rather US-centric assumptions about local governance that do not necessarily apply in many other contexts.

Again, the strong emphasis on internal institutionalisation, through new roles and offices, and on external institutionalisation, through setting up broader collaborations including the private sector, make the implementation of 100RC a relevant case for the study of the intersection between policy mobilities *through* TMNs and networked governance. We now move to present how two cities performed against certain broad dimensions that are key in contextualising our analysis.

8.3 Milan and Rome: Comparable Regulative Contexts but Quite Diverging Trajectories

Unsurprisingly, the two cities share some relevant contextual factors stemming from belonging to the same country and therefore the same multi-level governance system. These factors include a relatively mayor-centred local political system, limited fiscal autonomy, and strong dependence on state money transfers. Moreover, the two cities have weak metropolitan governance schemes (Del Fabbro 2018) and, due to long-standing austerity measures, also have variably weakened administrations in general in terms of human and budgeting resources, with Rome being in a particularly precarious position (Vicari and Violante 2018).

Within this broad context, they do present significant differences as well. Milan shows a pattern of relative local political stability with a major change in 2011—a shift from a centre-right to a centre-left mayor and majority—and then the confirmation of the same majority with a new mayor in 2016. Rome has a pattern of greater instability with a major change in 2013, a shift from a centre-right to centre-left mayor and majority, a caretaker government in 2015, and a new mayor and political majority—belonging to the 'populist' 5 Star movement in 2016—whose rule has been characterised by great instability. Also, the profiles of the actual mayors—Giuseppe Sala in Milan and Virginia Raggi in Rome—differ strongly. Sala—elected in 2016 as the candidate of the ruling centre-left coalition—has a record of management positions within the private sector, relevant international experience, and a key appointment as manager of the 2015 Expo in Milan. The younger Raggi—elected in 2016 as well—was a lawyer, then a 5 Star opposition city councillor and had no relevant international experience.

Looking at the wider, recent policy contexts, Milan has been praised as a successful model of urban policy characterised by a booming real estate market, international events, and economic and demographic growth (Andreotti 2019), while Rome has been depicted as a case of multidimensional urban crisis encompassing fiscal problems, corruption, failure in the provision of basic services, and a decline in real estate values (Coppola 2018b). In this context, Milan has a more consistent record of recent urban policy and planning initiatives, including successful large urban regeneration schemes (Pasqui 2018); a new master plan (2020) and a recent energy and environment plan (2018); and the launch of new initiatives in the areas of public space, public participation, and economic development with an overall clearer pattern of networked governance based on intensive collaborations with private foundations and companies, academic institutions, and civil society (Caudo and Coppola 2020).

Rome shows a less consistent pattern in the area of planning. Its last Master Plan was approved in 2008 with a limited ability to implement large urban regeneration programs (Baioni 2018; Busti 2018), there have been no recent major economic development initiatives and some limited attempts at public participation (Allegretti 2018). Overall, Rome has a weaker pattern of networked governance, as the role of foundations is way more limited than in Milan, collaboration with universities is episodic, and private–public partnerships are more traditionally understood (Coppola 2018a).

In this framework, we can observe differences in the ways the cities connected policy-making processes to international networks and collaborations. In Milan, city administrations heavily invested in international relationships beginning in 2011 conducting a major international exercise in the area of food policy related to the 2015 Expo event (The Food Policy Pact), strategically engaging in some TMNs (e.g. C40) as tools for branding and local policy-making leverage, strengthening participation in EU based projects (Urban Innovative Actions, H2020, Life), and promoting new collaborations such as the one with Bloomberg Associates (BA)—a new consulting philanthropy launched by the former mayor of New York City—and the participation to new international initiatives such as the C40 sponsored Reinventing Cities (Coppola and Citroni 2020). Mayors have also been investing in their personal participation in international events, establishing special relationships with some cities and taking part in high-level TMN sponsored global mayoral initiatives.

In Rome, after a relaunch of international activities between 2013 and 2015—involving the strengthening of participation in EU projects (H2020), the establishment of relationships with BA, UN Habitat, and some city-to-city diplomatic activity—there has been a relative decline with the administration in place since 2016, the only notable exception being the city participation in C40 and its 'Reinventing Cities' initiative. The degree of Rome's mayoral exposure in international events and networks appears to be considerably lower as well.

In general terms, as shown in the table below (Table 8.1), the international activities of Milan's city government seem to be significantly more sustained than Rome's.

The last aspect to address is the degree of institutionalisation of discourses around urban resilience. Both cities belong to a state where resilience policies, and more specifically climate resilience policies, are seldom institutionalised: the country has approved a climate adaptation strategy that appears to be scarcely operationalised, and the adoption of climate adaptation planning is not binding at the local level. The two cities, however, have significantly diverging patterns: Milan had a stronger record of policy activism with the establishment of some science-policy collaborations between local institutions (the city, i.e. the metropolitan area, and Politecnico di Milano on climate adaptation), an active role played by the Cariplo bank foundation in supporting and funding resilience initiatives at the regional and local levels, and finally the recent launch of a 'Climate and Air Climate' strategic initiative. Rome has a weaker pattern of recent environmental and energy planning (Nessi 2018) with limited science-policy collaborations and no significant and structured engagement on climate resilience with the participation of different actors (enterprises, philanthropies, etc.) (Coppola 2019b). Rome's Metropolitan Authority has been active in environmental policy and the issue of climate adaptation is mentioned in the strategic plan but no significant specific activity on climate adaptation policies is on the record.

Overall, based on this evidence, we can assume that international activities in Milan as opposed to Rome have been more widely and continuously conceptualised as relevant resources in mayors' and urban governments' legitimisation strategies and, more broadly speaking, in the workings of a city administration more active

Table 8.1 Participation in TMNs and EU project networks

	Milan	Rome
TMN	Eurocities 100 RC C40 (hosting network functions) City Protocol OSCE champion mayors for inclusive growth European Forum for Urban Security World Cities Culture Forum	100RC C40
EU project-based networks	Approach (Urbact) political inclusion of EU citizens Food for Cities (URBACT)—food policy Clever Cities (Horizon 2020)—Green Infrastructures Synchronicity (Horizon 2020) (Smart City) Dynamap (LIFE)—Noise Map Prepair (Life)—Air quality in the Po Valley Trifocal (Life)—food policy Veg-Gap (Life)—Air Quality Greening Fab employment and social innovation Fit food 2030—Patto sul Cibo Open Agri (UIA)—food policy Wish-Mi (UIA)—child poverty	Rurban (URBACT)—urban gardens Smart Mature Resilience (Horizon 2020) City Risks (Horizon 2020) Equal City—services for victims of sexual violence (Rights, Equality and Citizenship Programme) RoSaE (Europe for Citizens Programme)—international understanding

in policy innovation. We now move to analyse the implementation of the 100RC initiatives in the two cities. From a methodological point of view, the analysis is based on qualitative empirical evidence gathered between February 2014 and April 2020 including essentially three kinds of data: (a) relevant policy materials, including city strategic and network documents[1]; (b) participation in relevant events such as workshops, meetings, and public presentations; and (c) 12 in-depth interviews with stakeholders variably involved in the Rome and Milan initiatives, including former city commissioners, project coordinators, city staff, and 100RC staff (Fig. 8.2).[2]

---

[1] As the the Resilience Strategy of Milan was yet to be released at the time of writing the version that was analysed for the purposes of this contribution was an earlier draft. The final text of the Strategy could contain slightly different content.

[2] One of the authors was personally involved in the implementation of the programme in Rome as coordinator of the Roma Resiliente Initiative, which was the first phase of the city's participation in 100RC, concluded in January 2016 with the publication of the Preliminary Resilience Assessment.

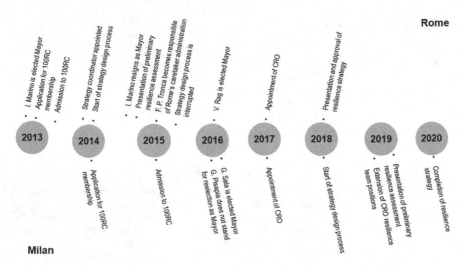

Fig. 8.2 Milestones and key events of Rome's and Milan's involvement with 100RC. *Source* Our data

In the following sections, we present and analyse the empirical evidence based on three analytical, partially overlapping dimensions: downscaling, internal institutionalisation, and external institutionalisation.

8.3.1 Downscaling

A first dimension we examined is the way in which the relationship was established between 100RC and the city administrations and their mayors, and the conflicts that arose in the negotiation of these relationships. We termed this dimension *downscaling*.

Starting from the genesis of the participation to the initiative, Rome was awarded participation in 100RC in 2013 based on a proposal prepared by the urban planning commissioner's office submitted with a supporting letter from the mayor. In contrast, Milan participated in the 2014 call with a proposal coordinated by the same office with the collaboration with Amat, a city environmental company, and Politecnico di Milano. The one-year offset made the application processes slightly more mayor-centred, as the 100RC management intended to ensure mayoral support for the initiative through in-person discussions (on-site or on-line) with the mayors of each candidate city (100RC staff and City Official Milan, Interviews 2020). Therefore, while in Rome a mayor's supporting letter was deemed sufficient, in Milan, a meeting (probably virtual) with the mayor apparently took place. In both cases, the rationales for the participation revolved around innovation in policy-making, internationalisation, the need to focus on climate and environmental issues in general and the access to the resources that 100RC entailed (100RC staff, Interview, 2020).

Looking at the main conflicts in both cities, but particularly in Milan, the main subject of negotiation between 100RC and the city rapidly became the institution-alisation of the new position and office. Reaching an agreement on this matter was the precondition for the actual kick-off of the process, and for the transfer of the funds aimed at essentially supporting the new position through a memorandum of understanding distributing the duties and responsibilities of the parties involved. In Rome, after one year of negotiations, a city-controlled agency linked to the urban planning office was identified as the organisation where the new, temporary position was to be located and a sizable working group created. In Milan, negotiations took longer both for the change of mayor that happened in 2016 and for the more pressing requests of 100RC, which asked for a clearer placement of the new position within the city administration and not within a city agency, as initially advocated by the city itself, and as close as possible to the mayor (100RC staff, Interview, 2020). This led to consistent delays and finally to the assignment of Milan's CRO as temporary director of a new office in 2018, almost three years after the city was included in 100RC.

Both cities were supported by 'city relationship managers' and 'platform part-ners': in the case of Rome, this was ICLEI—a large Germany-based TMN whose board included Rome's Deputy Mayor for the Environment at that time—while in the case of Milan it was ARUP that happened to have an office in Milan.

In the case of Rome, 100RC's decision to change the city partner from ICLEI to ARUP at the end of 2015 was a contentious issue that was resolved in favour of 100RC, as 100RC had in fact taken the decision to interrupt its relationship with ICLEI globally. Another key difference between Milan and Rome was that in Rome the resignation of the mayor in November 2015 led to the interruption of the strategy-design process, as the new caretaker administration did not renew its engagement with the organisation. The Raggi administration elected in 2016 re-established the relationships by appointing the city manager as CRO and completed the strategy process within one year.

Looking at the issue of the mayoral exposure and involvement we can observe that, in both cases, participation in the programme was supported by the mayors in place at the time although, as we have seen, the application processes were slightly different. However, in this regard there was variance both in terms of cities and in terms of mayors. In Rome's first phase, the mayor (Ignazio Marino) did not participate in any public events related to the strategy process because—the city commissioners in charge would participate in many events instead—and not even in a closed-door 'mayoral summit' event taking place in Italy. Later on, Mayor Virginia Raggi's expo-sure would significantly increase with her participation, along with many commis-sioners, in the presentation event for the final resilience strategy in 2018. In Milan, the mayor (Giuliano Pisapia) participated in the first workshop of the strategy process before the elections, while mayor Sala at this time has not participated in any relevant public event related to the strategy, but referred to it—and in broader terms to the city's participation to 100RC—on several occasions. In this case, it should be noted that the final Milan strategy had yet to be publicly presented—September 2020—due to the Covid-19 emergency. In both cities and for all mayors, however, closed-door meetings with 100RC staff were held, 'urban resilience' was mentioned in press issues and speeches, and the granting of the participation in the network was saluted as a significant achievement.

A final relevant issue is the negotiation of the very nature of the strategy, i.e. its policy breadth and relations with other relevant policies. On this issue, 100RC's initial, overly ambitious holistic and cross-sectoral approach seems to have evolved over time, becoming more and more realistic and able to tactically adapt to the two different local environments. At the same time, the alignment of the 'strategy partners' in the two cities—Rome changed from ICLEI to ARUP—seems to have ensured a higher degree of control and homogenisation of the two strategies. The discussion in the next two sections of the nature of the strategy and the breadth of the actors involved will further clarify this aspect.

8.3.2 Internal Institutionalisation

The second dimension we address is that of internal institutionalisation, which involves issues such as the design of the new position and office, the nature of the strategy design process, and its linkage to political responsibilities and priorities and the degree of legitimatisation across government and policy processes.

On the first issue, we have seen how the placement of the new CRO position within the city administration's hierarchies was the most relevant subject of negotiation and contention. In both cities, despite 100RC's expectations that the new position be very high-ranking and in direct contact with the mayor, it was accepted that it was not very high-ranking and put under the supervision of specific deputy mayors: in both cases CROs reported to the Deputy Mayor for urban planning, with a secondary role in Rome under Mayor Marino was reserved to the Deputy Mayor for the environment. In this latter case, as we have seen, there was no establishment of a new office but the formation of a sizable working group within a city-controlled agency. Taking aside a long period of inactivity, this choice was never repealed, while under the current mayor the position of the CRO became just a formal title for the overburdened city director. In the case of Milan, the position of the CRO was newly created within the city administration along with an office—formed by two new temporary positions and some city employees and trainees—initially located in the mayor's office, then moved to the planning office and finally relocated within a larger office dealing with the climate and environment (Rome City Official and 100RC staff, Interviews, 2020). Regarding long-term institutionalisation, despite being included in the 2019 strategy, the new 'resilience office' in Rome was never put in place, while in Milan the CRO position and office were extended after the end of the 100RC funding period up to the end of the current city legislature (2021).

Besides the issues with the new office's formal placement within city government, there is the substantial dimension of the day-to-day operations of the programme. Here the key issue is its perceived legitimacy and relevance across institution as a whole, as interviewees in both cities refer to issues such as the variability of the level of recognition and collaborative attitudes in different offices and agencies, fears of 'invasion' in established areas of policy by incumbent policy makers, obstacles to the new office's involvement in relevant policy processes, scepticism towards a new 'buzzword', and the overly 'process-oriented' and not very 'technical' skills of the staff involved in the process (100RC staff, Interview, 2020). Although these issues appear to have been relevant in both cities, in Milan it appears that they did not prevent the formal inclusion of the new office staff in several city government bodies and, as we will see, the exercise of some influence over established policy-making processes (Milan City Official and 100RC staff, Interviews, 2020). However, in both cities the urban resilience operations put in place were not in a position to take leadership in the management of crises whose nature was very consistent with the resilience discourse: traumatic events such as floods, heat waves, the migration crisis, and the Covid-19 crisis were mostly managed by established actors even if in the case of the latter Milan's resilience office was involved in the drafting of a "covid adaptation strategy".

Another dimension to look at in order to make sense of the processes of internal institutionalisation is the nature of the strategy design path. As we have seen, the process proposed by 100RC was highly structured and organised around certain milestones and certain tools. In Rome, at least during the first phase, the process was intense, highly structured, and aimed at the involvement of all areas of city government through a series of workshops focussing on the assesement of current conditions and the identification of priority areas. Evidence collected during other consultative processes—the planning office held a series of 'neighbourhood planning conferences' in the same period (Coppola 2017)—were presented as having informed the final document for this early phase as well (Roma Resiliente 2016). After the mentioned long interruption, the process restarted mostly in the form of internal meetings and interviews with deputy mayors and directors up to the final set-up of the strategic document, which was presented at a large event in the presence of the mayor in 2018 (Rome City Official, Interview, 2020). In Milan, a series of workshops were held involving a diversity of actors in the first phase and evidence was collected through other consultative processes taking place in the same period, i.e. those in connection to the new city master plan and to an initiative around the reopening of the city's historical canals (Milan City Official, Interview, 2020). The identification of the actions was mostly the outcome of an internal review process. In terms of the legislative significance, Rome approved the strategy with a city government resolution, while Milan has not yet published it. It is important to underline that the mentioned lack of institutionalisation of resilience strategies in the Italian framework makes its approval by political bodies such as city governments and councils a highly rhetorical exercise. However, while in Milan the linkage of certain actions identified in the strategies with binding policies and regulations was achieved with the change of certain rules in the new spatial masterplan, in Rome no significant outcomes in this sense seem to have been achieved.

8.3.3 External Institutionalisation

Finally, the third dimension we address is that of external institutionalisation, meaning the establishment of active relationships within the 100RC network itself—the strategy partner, the mentioned platform partners, and other cities—and the involvement of other actors both in the strategy design process and, more importantly, in the actions included in the final strategies.

In general, Rome was characterised as a less 'receptive' environment for exercises revolving around public–private partnerships and collaboration, a characterisation that was also based on a supposed suspicion towards the private sector (100RC staff, Interview, 2020). The relationship with the strategic partners was considered limited in this regard, as the first partner was the TMN ICLEI and the following partner, the Milan-based ARUP, was considered not sufficiently knowledgeable about Rome. In terms of city-to-city relationships, initial collaborations with some cities such as Rotterdam and New Orleans around issues of water management do not appear to have had a significant follow-up. As a result, Rome's strategy does not include actions involving any of these partners and is generally appreciably more limited in terms of setting up collaborations (see Table 8.2). In contrast, in Milan, ARUP played a key role (Milan City Official and 100RC staff, Interviews, 2020) and it became a partner in several strategy actions, as was the case with some 100RC platform partners. Milan also set up closer collaborations with some cities aimed at developing certain actions by translating relevant practices from their respective settings: Manchester and Rotterdam were mentioned as important examples sources

Table 8.2 Distribution of actors by type in actions included in the resilience strategies of Milan and Rome

Action	Rome	Milan
100RC	–	2
100RC partners	–	3
Other levels of government (national, regional, metropolitan area)	9	2
National agencies/institutes	2	1
Municipal and state-owned companies	3	6
Private companies	12 (generic)	7
Universities and research institutions	3	9
Foundations	–	5
Associations	17 (generic)	5
Civil movements	–	2
TMNs (other than 100RC)	–	1
Other cities	–	4
Supranational organisations	1	–
Actions without collaborating partners (city administration only)	13	3

in an urban forestation project (see below), while Paris and Barcelona were mentioned as important for setting up an initiative on school gardening (Milan City Official, Interview, 2020). From the perspective of 100RC's staff, these diverging experiences revealed the very different abilities of the two local contexts to make policy innovations and to be 'champions' of the resilience discourse by providing an effective example to be imitated around the world (100RC staff, Interview, 2020).

A second issue in this respect is the overall breadth of the involvement of actors other than the city administration in the strategy design and implementation processes (see again Table 8.2). As mentioned, both cities have promoted consultative processes (Roma Resiliente 2016) that have involved actors from civil society and business: Rome held workshops expressly defined to include business, civil society, and social activism actors in the initial phase (Roma Resiliente 2016), as did Milan, although in less structured ways (100RC staff, Interview, 2020). Moving to the strategy documents, substantial and clear differences make themselves evident. Milan's strategy has on average a higher proportion of actions involving multiple actors, and those actors are specifically named, while Rome has fewer actions of this kind and often no clear indication of the identity of the actors involved is offered; only broad families of actors are mentioned, such as 'associations' or 'private partners'. In this wider 'networkedness' of Milan's strategy, of particular significance is the strong presence of the Cariplo Foundation, of local academic institutions—especially Politecnico di Milano, which participates in many of the city's EU-funded projects—and of other cities. These substantial differences are clearly linked to the already mention different nature of the two strategies. In Rome, lacking other significant strategies, the resilience strategy was initially conceived as a wide scope strategy aimed at identifying new policies and actions. With the new administration it was turned mostly into a process aimed at restating of priorities the mayor had already indicated in her campaign manifesto (Rome City Official, Interview, 2020). In Milan, the strategy was conceived instead as 'a collection of projects' because, in the words of city staff, 'governing a city is probably more likely to happen through single projects and not overarching, conforming plans' (Interview with City Official, April 2020). Many of these projects were actions already in the making that appeared to have significance in a 'resilience perspective', while others were policy and regulative innovations in limited areas—i.e. the master plan—or new initiatives presented as 'flagship' interventions. One initiative in particular, named Forestami and aimed at increasing tree planting in metropolitan areas, appears to be the poster child for this networking governance ethos, as it is based on the participation of two private foundations—one linked to a major actor involved in the most significant redevelopment plan in the city—and Politecnico di Milano based on a scheme that, as mentioned, was created with the cooperation of other cities in the 100RC network (Milan City Official and 100RC staff, Interview, 2020).

8.4 Discussion

As we have seen, 100RC has been an overly ambitious initiative aimed at spreading a discourse, institutionalising a new position and new policy tools within city governments. Unlike other TMNs, this ambition and the fuzziness of its thematic focus made the exercise particularly difficult to operationalise but also adaptable to different contexts. In this regard, Milan and Rome shared a framework in which such an exercise could not be based on any formal, higher scale policy mandates such as a climate adaptation strategy and in which discourses around urban resilience—in Rome especially—had yet to become very influential. In this context, looking at the overall official implementation record of the initiative in the two cities differences may not seem particularly relevant: both cities experienced changes in administrations that entailed significant delays in their strategy design process and some friction regarding the definition of its essential organisational terms; and both cities managed to complete the strategy design process with both mayors ensuring a certain degree of exposure and visibility to the resilience strategies.

Nevertheless, by looking more in detail at the dynamics related to the three analytical dimensions identified, we can find a significantly diverging pattern in critical areas such as the setting up the new office and its legitimisation within the city government, the nature of the strategy and its design process, the degree of collaboration with members of the 100RC network, and the degree of inclusion of external local and non-local actors. These differences seem to depend, at least in part, on the local government and governance contexts and on the different mayors' and urban governments' legitimisation strategies. The Mayor of Milan—and the city administration at large—seems to have a more 'policy-based' legitimisation strategy that can also leverage a previous legacy of strong engagement in international activities and EU funding, which are seen as important factors in strengthening networked governance mechanisms and policy innovations. In Rome, however, the nature of the mayor's legitimisation strategy seems less clear, largely devoid of a legacy or context of that kind and therefore seeming to point to a form of 'rhetorical appropriation' of the resilience initiative as an opportunity to reinstate more generic political electoral priorities.

In this context, we can observe how the core of the 100RC initiative—i.e. the institutionalisation of a high-ranking position—was in both cases a matter of contention but with different outcomes. In Rome, while institutionalisation was formally included in the final strategy, it does not seem to have ever become a real option, as the provision of the project was externalised to a city-controlled agency, and the appointment of a CRO in the second phase was mostly rhetorical. In Milan, however, there were significant steps in the direction of true institutionalisation, with the creation of a new office within the administration and the extension of its duration up to the end of the legislature and therefore beyond the closure of the 100RC initiative. This is certainly also related to the time offset between the two cities' participation in the program—one that allowed 100RC to elaborate a clearer strategy in that regard in Milan—but the fact that the reprisal of the process in Rome after a long interruption did not lead to comparable outcomes is indicative of a deeper divergence. What is instead shared by the two cities, and therefore significant in terms of the limitations of the initiative, is the fact that the strategy design process was made the responsibility of deputy mayors to signal the indirect involvement of mayors and the need to operationalise a fuzzy understanding of resilience through its placement within a specific specialised area of government.

Also, the impetus towards 'collaborative' ways of governance, particularly public–private partnerships, led to very different outcomes. As we have seen, one key difference between Milan's and Rome's strategies is the inclusion in the former case of a larger variety of clearly identified actors, very often external to the realms of public governance, that seem instead to be central in the latter case. In Milan, the 'projects-based strategy' based on recombining existing initiatives to be mainstreamed as resilience initiatives, some innovations in existing policy frameworks, and the adoption of new initiatives among actors already involved in local governance meant that many partners included in certain actions were already present in other policy arenas. In this sense, the resilience strategy was one more occasion for these actors to reproduce and further legitimise their participation in this variety of policy arenas and reinforce a wider local governance ethos based on ideas of collaboration, leveraging on existing actions and actors and on 'public policies that do not impose actions but enable actors' (City Official, Interview, 2020). This approach seems to have met with 100RC's favour, as it was increasingly interested in delivering concrete results rather than having strategies resembling a 'doctoral thesis' (100RC staff, Interview, 2020). In Rome, besides an initial phase of wider 'stakeholder engagement' in 2015, the strategy process appears to have been decisively internalised in 2019, and this took place in an overall context featuring fewer governance networks, less collaboration, and a more traditional pattern of relationships between the city administration and established interests.

Within this context, the style of action of CROs and the wider staff also seem to be quite different, even if both cases encountered the usual obstacles such as defensive behaviours and scepticism of buzzwords on the part of established political and administrative actors. In Milan, the role of the CRO seems to be in some way that of a 'policy entrepreneur' (Mintrom and Norman 2009) capable of navigating existing collaboration opportunities and entering into a limited number of policy processes by proving the relevance of certain resources channelled from the 'network', such as relationships with other cities and the mobilisation of 100RC partners. In Rome, after a first phase in which there seems to have been a comparable logic of action, the exercise in the second phase seems to have been directly appropriated by political figures in the form of an internal exercise in coordination between a disempowered working group, the mayor, and the other members of the administration's staff.

8.5 Conclusions

This contribution asked what happens when a global-scale and apparently one-size-fits-all approach to building urban resilience encounters the reality of variegated local contexts. In this regard, in contrast to most of the literature on policy mobilities, the case examined seems to be centred on the mobility of a discourse, frameworks, and governance models rather than, as traditionally understood, the mobility of policies per se. Also, the relational structure within which this mobility was produced was peculiar, pushing to partially reframe the literature's common emphasis on movement and adaptation.

Rather than a classic TMN, 100RC was a highly centralised organisation based on the formalisation of network-like relationships with a wealth of private actors and on the selection—not the adhesion—of member cities. By creating a breed of 'urban resilience champions'—cities and leaders—that could 'lead by example on a global stage', 100RC pursued the mainstreaming of a certain discourse on urban resilience by using mayors as vehicles of a governance reform agenda built on ideas of innovations in government and partnerships with private and civil society actors. Offering diagnostic tools to reframe existing urban problems, establishing a whole new field of relations among cities, large consultancies, and companies, the initiative aimed to besiege a supposed 'old way' of governing characterised by the fragmentation of public action into policy silos and a lack of collaboration with resourceful local and global actors.

In this framework, selected cities were exposed to a particularly intense set of expectations, pressures, and requirements. But as actual engagement processes unfolded, an apparently one-size-fits-all approach encountered varying mayoral political legitimation strategies and different local and institutional contexts characterised by changing degrees of penetration of networked governance models. Initial assumptions had to be reframed, goals had to be adjusted, and cities had to be constructed by the initiative as differently 'receptive' environments, to which city-management strategies had to be adapted and tailored. If in some of these contexts the most to be obtained was the completion of highly rhetorical exercises, other, more 'receptive' contexts could be constructed as potential 'champions' that could more intensively use the resources made available by the organisation.

The case discussed seems to point to a policy mobility model that is both "diffusive" and peculiarly based on the circulation of policy frameworks and tools and that, as described in this contribution, embeds very differently in different cities. This suggests that one-size-fits-all approaches must in any case be implicitly or explicitly negotiated under a variety of local conditions, and that scholars increasingly need to be able to disentangle the subtle, evolving, and multi-located motives that push cities, organisations, and networks to engage in such challenging exercises.

Acknowledgements We are very grateful to Prof. Alberta Andreotti of Università di Milano Bicocca and Prof. Alberto Vanolo of Università di Torino for their comments and insights that greatly helped us to better develop this contribution.

References

100RC (undated) 100 Resilient Cities—Pioneered by the Rockefeller Foundation. Retrieved from https://www.100resilientcities.org/

Allegretti G (2018) Cercando radici di futuro? Recupero di memoria di un sistema partecipativo incompleto, In: Coppola A, Punziano G (eds), Roma in transizione: governance, strategie, metabolismi e quadri di vita di una metropoli. Planum Publisher, Roma-Milano

Andreotti A (2019) Governare Milano nel nuovo millennio. Il Mulino, Bologna

Baioni M (2018) Le compensazioni a Roma: da espediente a ipoteca sul futuro. In: Coppola A and Punziano, Roma in Transizione. Governo, strategie, metabolismi e quadri di vita di una metropoli, vol 1. Planum Publisher, Roma-Milano

Baravikova A, Coppola A, Terenzi A (2020) Operationalizing urban resilience: insights from the science-policy interface in the European Union. Eur Plann Stud. https://doi.org/10.1080/096 54313.2020.1729346

Barber B (2013) If mayors ruled the world: dysfunctional nations. Yale University Press, New Haven, CT

Berkowitz M (2015) What is a CRO and what are they doing in Mexico city? Retrieved from https://www.100resilientcities.org/what-is-a-cro-what-are-they-doing-in-mexico-city/

Bogason P, Musso JA (2006) The democratic prospects of network governance. Am Rev Publ Adm 36(1):3–18. https://doi.org/10.1177/0275074005282581

Brenner N (2004) New state spaces: Urban governance and the rescaling of statehood. Oxford University Press, Oxford

Bulkeley H, Davies A, Evans B, Gibbs D, Kern K, Theobald K (2003) Environmental governance and transnational municipal networks in Europe. J Environ Plann Pol Manage 5(3):235–254. https://doi.org/10.1080/1523908032000154179

Busch H (2015) Linked for action? an analysis of transnational municipal climate networks in Germany. Int J Urban Sustain Dev 7(2):213–231. https://doi.org/10.1080/19463138.2015.105 7144

Busti M (2018) Le centralità di Roma: strategia e prassi. In: Coppola A, Punziano G (eds) Roma in Transizione. Governo, strategie, metabolismi e quadri di vita di una metropoli, vol 1. Planum Publisher, Roma-Milano

Caudo G, Coppola A (2020) Orizzonti di innovazione democratica. In: Urban@It, Quinto Rapporto sulle città. Il Mulino, Bologna

Coaffee J (2013a) Towards next-generation urban resilience in planning practice: from securization to integrated place making. Plann Pract Re 28:323–339. https://doi.org/10.1080/02697459.2013. 787693

Coaffee J (2013b) Rescaling and responsibilising the politics of urban resilience: from national security to local place-making. Politics 33(4):240–252. https://doi.org/10.1111/1467-9256.12011

Coaffee J (2016) Terrorism, risk and the global city: towards urban resilience. Routledge, London

Coppola A (2017) Roma prossima. Territorio 82:68–71. https://doi.org/10.3280/TR2017-082015

Coppola A (2018a) Roma in Transizione. Problemi pubblici emergenti fra scienza ed azione pubblica. La sfida (rimandata?) delle governance e delle politiche della complessità a Roma. In: Coppola A and Punziano G (eds) Roma in Transizione. Governo, strategie, metabolismi e quadri di vita di una metropoli, vol 2. Planum Publisher, Roma-Milano

Coppola A (2018b) Roma in Transizione. Studiare Roma in Transizione. Temi e problemi. In: Coppola A and Punziano (eds) Roma in Transizione. Governo, strategie, metabolismi e quadri di vita di una metropoli, vol. 2. Planum Publisher, Roma-Milano

Coppola A, Citroni S (2020) The emerging civil society. Governing through leisure activism in Milan, Leisure Studies https://doi.org/10.1080/02614367.2020.1795228

Cox KR (1993) The local and the global in the new urban politics: a critical view. Environ Plann D: Soc Space 11(4):433–448. https://doi.org/10.1068/d110433

Crivello S (2015) Urban policy mobilities: the case of Turin as a smart city. Eur Plann Stud 23(5):909–921. https://doi.org/10.1080/09654313.2014.891568

Croese S, Green C, Morgan G (2020) Localizing the sustainable development goals through the lens of urban resilience: lessons and learnings from 100 resilient cities and cape town. Sustainability 12(2):550. https://doi.org/10.3390/su12020550

Davies J (2011) Challenging governance theory: from networks to hegemony. Bristol University Press

Davoudi S, Porter L (2012) Applying the resilience perspective to planning: critical thoughts from theory and practice. Plann Theory Pract 13(2):299–333. https://doi.org/10.1080/14649357.2012.677124

Del Fabbro M (2018) The institutional history of Milan metropolitan area. Territ Polit Govern 6(3):342–361. https://doi.org/10.1080/21622671.2017.1369895

Dircke P, Molenaar A (2015) Climate change adaptation; innovative tools and strategies in Delta City Rotterdam. Water Pract Tech 10(4):674–680. https://doi.org/10.2166/wpt.2015.080

Evans JP (2011) Resilience, ecology and adaptation in the experimental city. Trans Inst Brit Geograph 36(2):223–237. https://doi.org/10.1111/j.1475-5661.2010.00420.x

Fastenrath S, Coenen L, Davidson K (2019) Urban resilience in action: the resilient melbourne strategy as transformative urban innovation policy? Sustainability 11(3):693. https://doi.org/10.3390/su11030693

Feldman DL (2012) The future of environmental networks—governance and civil society in a global context. Futures 44(9):787–796. https://doi.org/10.1016/j.futures.2012.07.007

Fenton P, Busch H (2016) Identifying the "usual suspects"—assessing patterns of representation in local environmental initiatives. Chall Sustainab 4(2):1–4. https://doi.org/10.12924/cis2016.04020001

Fisher S (2014) Exploring nascent climate policies in Indian cities: a role for policy mobilities? Int J Urban Sustain Dev 6(2):154–173. https://doi.org/10.1080/19463138.2014.892006

Hakelberg L (2014) Governance by diffusion: transnational municipal networks and the spread of local climate strategies in Europe. Global Environ Polit 14(1):107–129. https://doi.org/10.1162/GLEP_a_00216

Hall T, Hubbard P (1996) The entrepreneurial city: new urban politics, new urban geographies? Prog Hum Geogr 20(2):153–174. https://doi.org/10.1177/030913259602000201

Harvey D (1989) From managerialism to entrepreneurialism: the transformation in urban governance in late capitalism. Geografiska Annaler Series B, Human Geogr 71(1):3. https://doi.org/10.2307/490503

Haupt W (2018) European municipalities engaging in climate change mitigation and adaptation networks: examining the case of the covenant of mayors. In: Yamagata Y, Sharifi A. (eds) Resilience-oriented urban planning: theoretical and empirical insights. Cham, Springer, pp 93–110. https://doi.org/10.1007/978-3-319-75798-8_5

Haupt W, Chelleri L, Van Herk S, Zevenbergen C (2019) City-to-city learning within climate city networks: definition, significance, and challenges from a global perspective. Int J Urban Sustain Dev. https://doi.org/10.1080/19463138.2019.1691007

Haupt W, Coppola A (2019) Climate governance in transnational municipal networks: advancing a potential agenda for analysis and typology. Int J Urban Sustain Dev 11(2):123–140. https://doi.org/10.1080/19463138.2019.1583235

Heynen N, Robbins P (2005) The neoliberalization of nature: governance, privatization, enclosure and valuation. Capit Na Soc 16(1):5–8. https://doi.org/10.1080/1045575052000335339

Jabareen Y (2013) Planning the resilient city: concepts and strategies for coping with climate change and environmental risk. Cities 31:220–229. https://doi.org/10.1016/j.cities.2012.05.004

Kern K, Bulkeley H (2009) Cities, Europeanization and multi-level governance: governing climate change through transnational municipal networks. J Common Market Stud 47(2):309–332. https://doi.org/10.1111/j.1468-5965.2009.00806.x

Khan J (2013) What role for network governance in urban low carbon transitions? J Clean Prod 50:133–139. https://doi.org/10.1016/j.jclepro.2012.11.045

Komninos N, Kakderi C, Panori A, Tsarchopoulos P (2018) Smart city planning from an evolutionary perspective. J Urban Tech 26(2):3–20. https://doi.org/10.1080/10630732.2018.148 5368

Leitner L, Sheppard E, Webber S, Colven E (2018) Globalizing urban resilience. Urban Geogr 39(8):1276–1284. https://doi.org/10.1080/02723638.2018.1446870

Makhoul N (2018) Seismic loss estimation of Byblos city: a contribution to the '100 Resilient Cities' Strategy. In: 16th European Conference on Earthquake Engineering. Thessaloniki

Marsh D, Sharman JC (2009) Policy diffusion and policy transfer. Pol Stud 30(3):269–288. https://doi.org/10.1080/01442870902863851

McCann EJ (2011a) Urban policy mobilities and global circuits of knowledge: toward a research agenda. Ann Assoc Am Geogr 101(1):107–130. https://doi.org/10.1080/00045608.2010.520219

McCann E (2011b) Veritable inventions: cities, policy and assemblage. Area 43(2):143–147. https://doi.org/10.1111/j.1475-4762.2011.01011.x

McEvoy D, Fünfgeld H, Bosomworth K (2013) Resilience and climate change adaptation: the importance of framing. Plann Pract Res 28(3):280–293. https://doi.org/10.1080/02697459.2013.787710

Mehmood A (2016) Of resilient places: planning for urban resilience. Eur Plann Stud 24(2):407–419. https://doi.org/10.1080/09654313.2015.1082980

Meerow S, Newell JP, Stults M (2016) Defining urban resilience: a review. Landsc Urban Plann 147:38–49. https://doi.org/10.1016/j.landurbplan.2015.11.011

Mcjía-Dugand S, Kanda W, Hjelm O (2016) Analyzing international city networks for sustainability: a study of five major Swedish cities. J Clean Prod 134:61–69. https://doi.org/10.1016/j.jclepro.2015.09.093

Melkunaite L, Guay F (2016) When civil protection meets urban planning: conceptualising a resilient city development process. WIT Trans Ecol Environ 204:455–466

Mell I, Allin S, Reimer M, Wilker J (2017) Strategic green infrastructure planning in Germany and the UK: a transnational evaluation of the evolution of urban greening policy and practice. International Planning Studies 22(4):333–349. https://doi.org/10.1080/13563475.2017.1291334

Mintrom M, Norman P (2009) Policy entrepreneurship and policy change. Pol Stud J 37(4):649–667. https://doi.org/10.1111/j.1541-0072.2009.00329.x

Nagorny Koring N (2019) Leading the way with examples and ideas? Governing climate change in German municipalities through best practices. J Environ Plann Pol Manag 21(1):46–60. https://doi.org/10.1080/1523908X.2018.1461003

Nessi H (2018) Mobilità dei modelli e resistenze del locale: Il caso masterplan di Jeremy Rifkin a Roma e il fallimento della transizione energetica, in Coppola A and Punziano, Roma in Transizione. Governo, strategie, metabolismi e quadri di vita di una metropoli, vol 2. Planum Publisher, Roma-Milano

Pasqui G (2018) Raccontare Milano. Politiche, progetti, immaginari. FrancoAngeli, Milano

Peck J, Theodore N (2010) Mobilizing policy: Models, methods, and mutations. Geoforum 41(2):169–174. https://doi.org/10.1016/j.geoforum.2010.01.002

Pierre JB, Peters G (2000) Governance. New York St. Martin's Press, politics and the state. https://doi.org/10.1111/j.1467-7660.1994.tb00519.x

Owens S, Petts J, Bulkeley H (2006) Boundary work: knowledge, policy, and the urban environment. Environ Plann C: Govern Pol 24(5):633–643. https://doi.org/10.1068/c0606j

Pitidis V, Tapete D, Coaffee J, Kapetas L, Porto de Albuquerque J (2018) Understanding the implementation challenges of urban resilience policies: investigating the influence of urban geological risk in Thessaloniki. Greece. Sustainability 10(10):3573. https://doi.org/10.3390/su10103573

Prince RJ (2010) Policy transfer as policy assemblage: making policy for the creative industries in New Zealand. Environ Plann 42(1):169–186. https://doi.org/10.1068/a4224

Restemeyer B, Woltjer J, Van den Brink M (2015) A strategy-based. Framework for assessing the flood resilience of cities – A Hamburg case study. Planning Theory & Practice 16:45–62. https://doi.org/10.1080/14649357.2014.1000950

Rodin J (2014) The resilience dividend: being strong in a world where things go wrong. Public Affairs, New York

Rosato V, Di Pietro, A La Porta L, Lavalle L, Pollino M, Tofani A (2018) Sistemi di supporto alle decisioni per l'analisi del rischio nelle aree metropolitane: uno sviluppo in corso sull'area di Roma Capitale, in Coppola A and Punziano, Roma in Transizione. Governo, strategie, metabolismi e quadri di vita di una metropoli, vol. 2. Planum Publisher, Roma-Milano

Robinson J (2015) Arriving At' Urban policies: the topological spaces of urban policy mobility. Int J Urban Reg Res 39(4):831–834. https://doi.org/10.1111/1468-2427.12255

Sharifi A, Chelleri L, Fox-Lent C, Grafakos S, Pathak M, Olazabal M, Moloney S, Yumagulova L, Yamagata J (2017) Conceptualizing dimensions and characteristics of urban resilience: insights from a co-design process. Sustainability 9(6):1032. https://doi.org/10.3390/su9061032

Shaw K, Theobald K (2011) Resilient local government and climate change interventions in the UK. Local Environ 16(1):1–15. https://doi.org/10.1080/13549839.2010.544296

Sørensen E (2014) Democratic theory and network governance. Adm Theory Praxis 24:693–720. https://doi.org/10.1080/10841806.2002.11029383

Spaans M, Waterhout B (2016) Building up resilience in cities worldwide—Rotterdam as participant in the 100 resilient cities programme. Cities 61:109–116. https://doi.org/10.1016/j.cities.2016.05.011

Stone D (2004) Transfer agents and global networks in the 'transnationalization' of policy. J Eur Publ Pol 11(3):545–566. https://doi.org/10.1080/13501760410001694291

Sutherland C, Roberts DC, Douwes J (2019) Constructing resilience at three scales: the 100 resilient cities programme, Durban's resilience journey and water resilience in the Palmiet Catchment. Human Geogr 34(12):33–49. https://doi.org/10.1177/194277861901200103

Svitková K (2018). Making a 'Resilient Santiago': private sector and urban governance in Chile. Czech Soc Rev 54(6):933–960. https://doi.org/10.13060/00380288.2018.54.6.436

Tenemos C, Baker T, Cook IR (2019) Inside mobile urbanism: Cities and policy mobilities. In: Schwanen T, van Kempen R (eds) Handbook of urban geography. Stockport, Edward Elgar 1:103–118. https://doi.org/10.4337/9781785364600

Vale LJ, Campanella TJ (2005) Introduction. The cities rise again. In: Vale LJ, Campanella TJ, The resilient city: How modern cities recover from disaster. Oxford University Press, Oxford

Vicari P, Violante A (2018) Ancora un'eccezione? La crisi fiscale della capitale tra neoliberismo e presunto declino. In: Coppola, A. and Punziano, G (eds) Roma in Transizione. Governo, strategie, metabolismi e quadri di vita di una metropoli, vol 1. Planum Publisher, Roma-Milano

Viitanen J, Kingston R (2014) Smart cities and green growth: outsourcing democratic and environmental resilience to the global technology sector. Environ Plann 46(4):803–819. https://doi.org/10.1068/a46242

Welsh M (2014) Resilience and responsibility: governing uncertainty in a complex world. Geogr J 180(1):15–26. https://doi.org/10.1111/geoj.12012

Printed in the United States
By Bookmasters